Springer Complexity

Springer Complexity is an interdisciplinary program publishing the best research and academic-level teaching on both fundamental and applied aspects of complex systems – cutting across all traditional disciplines of the natural and life sciences, engineering, economics, medicine, neuroscience, social and computer science.

Complex Systems are systems that comprise many interacting parts with the ability to generate a new quality of macroscopic collective behavior the manifestations of which are the spontaneous formation of distinctive temporal, spatial or functional structures. Models of such systems can be successfully mapped onto quite diverse "real-life" situations like the climate, the coherent emission of light from lasers, chemical reaction-diffusion systems, biological cellular networks, the dynamics of stock markets and of the internet, earthquake statistics and prediction, freeway traffic, the human brain, or the formation of opinions in social systems, to name just some of the popular applications.

Although their scope and methodologies overlap somewhat, one can distinguish the following main concepts and tools: self-organization, nonlinear dynamics, synergetics, turbulence, dynamical systems, catastrophes, instabilities, stochastic processes, chaos, graphs and networks, cellular automata, adaptive systems, genetic algorithms and computational intelligence.

The two major book publication platforms of the Springer Complexity program are the monograph series "Understanding Complex Systems" focusing on the various applications of complexity, and the "Springer Series in Synergetics", which is devoted to the quantitative theoretical and methodological foundations. In addition to the books in these two core series, the program also incorporates individual titles ranging from textbooks to major reference works.

Editorial and Programme Advisory Board

Péter Érdi
Center for Complex Systems Studies, Kalamazoo College, USA and Hungarian Academy of Sciences, Budapest, Hungary

Karl Friston
Institute of Cognitive Neuroscience, University College London, London, UK

Hermann Haken
Center of Synergetics, University of Stuttgart, Stuttgart, Germany

Janusz Kacprzyk
System Research, Polish Academy of Sciences, Warsaw, Poland

Scott Kelso
Center for Complex Systems and Brain Sciences, Florida Atlantic University, Boca Raton, USA

Jürgen Kurths
Nonlinear Dynamics Group, University of Potsdam, Potsdam, Germany

Linda Reichl
Center for Complex Quantum Systems, University of Texas, Austin, USA

Peter Schuster
Theoretical Chemistry and Structural Biology, University of Vienna, Vienna, Austria

Frank Schweitzer
System Design, ETH Zurich, Zurich, Switzerland

Didier Sornette
Entrepreneurial Risk, ETH Zurich, Zurich, Switzerland

Understanding Complex Systems

Founding Editor: J.A. Scott Kelso

Future scientific and technological developments in many fields will necessarily depend upon coming to grips with complex systems. Such systems are complex in both their composition – typically many different kinds of components interacting simultaneously and nonlinearly with each other and their environments on multiple levels – and in the rich diversity of behavior of which they are capable.

The Springer Series in Understanding Complex Systems series (UCS) promotes new strategies and paradigms for understanding and realizing applications of complex systems research in a wide variety of fields and endeavors. UCS is explicitly transdisciplinary. It has three main goals: First, to elaborate the concepts, methods and tools of complex systems at all levels of description and in all scientific fields, especially newly emerging areas within the life, social, behavioral, economic, neuro- and cognitive sciences (and derivatives thereof); second, to encourage novel applications of these ideas in various fields of engineering and computation such as robotics, nano-technology and informatics; third, to provide a single forum within which commonalities and differences in the workings of complex systems may be discerned, hence leading to deeper insight and understanding.

UCS will publish monographs, lecture notes and selected edited contributions aimed at communicating new findings to a large multidisciplinary audience.

For other titles published in this series, go to
www.springer.com/series/5394

Syamal K. Dana · Prodyot K. Roy ·
Jürgen Kurths
Editors

Complex Dynamics in Physiological Systems: From Heart to Brain

Springer

Editors
Dr. Syamal K. Dana
Central Instrumentation
Indian Institute of Chemical
Biology
Kolkata-700032
India
skdana@iicb.res.in

Dr. Prodyot K. Roy
Department of Physics
Presidency College
Kolkata-700073
India

Dr. Jürgen Kurths
Institut für Physik
Universität Potsdam
Am Neuen Palais 10
14415 Potsdam
Germany
juergen@gmx.de

ISBN: 978-1-4020-9142-1   e-ISBN: 978-1-4020-9143-8

DOI:10.1007/978-1-4020-9143-8

Library of Congress Control Number: 2008935163

© Springer Science+Business Media B.V. 2009
No part of this work may be reproduced, stored in a retrieval system, or transmitted
in any form or by any means, electronic, mechanical, photocopying, microfilming, recording
or otherwise, without written permission from the Publisher, with the exception
of any material supplied specifically for the purpose of being entered
and executed on a computer system, for exclusive use by the purchaser of the work.

Printed on acid-free paper

9 8 7 6 5 4 3 2 1

springer.com

Preface

This book is an outcome of the *International Workshop on Complex Dynamics of Physiological Systems* held in the Department of Physics, Presidency College, Kolkata, India in February, 2007 and participated by about 100 researchers from different field, physics, mathematics, biology, and medicine. The workshop has a special significance due to its unique location at the oldest Indian college, which was founded in the year 1817 as Hindu College later renamed as Presidency College in 1855. The college has the distinction of being the institution where Nobel Laureate Professor Amartya Sen (Economics, 1998), Professor S. N. Bose (Bose-Einstein Statistics) and Oscar winner Satyajit Ray (Lifetime Achievement, 1992) studied.

The workshop was organized to review main directions and perspectives of chaos synchronization and its applicability in medical science and physiology, particularly, heart and brain. Nonlinear dynamics has become an important field of research since 1990 after the invention of phenomena and methods of controlling and synchronizing chaotic dynamics. Especially, application of chaos and synchronization appear highly promising in the field of biology, particularly, in physiological systems like heart and brain. However, several fundamental issues of complex dynamics, in particular, in spatiotemporal systems are yet to be resolved for its expansion to such applications. Analyzing corresponding multivariate time series from experimental measurements to extract information about a system's behavior and its prediction are the focal issues in this direction. Modeling complex systems like the heart and the brain still remains a challenging task. Understanding the mechanisms of cognitive response of human being to real world events is at the top of the agenda. Finally, to develop suitable diagnostic techniques based on nonlinear dynamics' tools using accessible measurements from such systems is important to address the hindrances in the way of their feasibility. Many researchers have addressed the issues recently, although a real progress is yet to be made.

Meanwhile, studies on the nonlinear dynamics of heart and brain have also started in India but they are mostly restricted to theoretical investigations of models and yet a few attempts are made on analyzing physiological data. Many groups in India are already involved in studies of nonlinear dynamical systems, in general. However, there is a missing link between physicists, mathematicians and physicians around the world, and India in particular, which needs to be bridged to unravel the complexities of physiological systems and to propose any suitable diagnostic tool at the clinical level. The workshop attempted to attract and to motivate young researchers, particularly, within India to investigations related to complex behavior in physiological systems and thereby initiate a future trend of research in complex dynamics of heart and brain using clinical data.

For this, the workshop was a great opportunity to bring together experts from various disciplines as physics, mathematics and cardiology and neuroscience within India and from abroad to discuss emerging issues on the field. A series of invited talks were presented by experts from different countries who are currently involved in studies of complexity in physiological systems, mainly, heart and brain and trying to explore potential applications of nonlinear dynamics in such systems. The scope of this book thus remains focused to the title of the workshop. In this endeavor, the articles are selected after peer review from the presented talks for inclusion in this book for a future reference.

The book consists of four chapters. The first one includes topics on data analysis. It deals with techniques how to reconstruct a strange attractor from a set of measured data, in general, how to extract information from measured electrophysiological signals like EEG and ECG, and to develop an understanding on the sources of heart rate variability from cardio-respiratory data. The second chapter is based on cardiovascular physics and modeling. This chapter discusses different cardiac models, origin of spiral chaos and turbulence and methods how to control or suppress them. Studies on heart models are presented to show how defects in heart muscle develop disturbance that spreads in excitable medium like heart to produce turbulence and thereby terminates into heart failure. The third chapter deals with the analysis of real data measured from the heart under healthy and pathological conditions and thereby to devise procedures for early prediction of heart disease. The chapter includes articles on synchronization of the cardio-respiratory system, and its relation to healthy and pathological situations. The fourth and final chapter focuses on neuroscience and cognitive behaviors. This chapter presents some of the issues like brain dynamics and epilepsy, behaviors of cell membrane, ions and channels under noisy conditions. It also includes cognitive behaviors like speech rhythms in children learning, theoretical understanding of the mechanisms of hearing and its instabilities, neuro-anatomical and chemical basis of *Drosophila* courtship behavior.

This book gives an overview for the interested readers in the emerging field of dynamics in physiological systems and its applications. Several articles even addressed to future directions on the topic of the workshop.

Finally, we would like to acknowledge the financial support forwarded by the Centre for Applied Mathematics and Computational Science, Saha Institute of Nuclear Physics (SINP), Kolkata, West Bengal University of Technology (WBUT), Kolkata, and Department of Science and Technology (DST) and Department of Atomic Energy (DAE), Council of Scientific and industrial Research (CSIR), Government of India, and Institute for Plasma Research (IPR), Gandhinagar, India, who came forward to encourage such a meeting of interdisciplinary nature. Not the least, we appreciate Prof. Mamata Roy, Principal, Presidency College, Kolkata, Prof. Ashoke Ranajan Thankur, Vice Chancellor, WBUT and Prof. Bikas K. Chakrabarti, SINP and Prof. Kamalesh Bhaumick, WBUT for their constant support and encouragements in organizing a meeting of new perspectives. We thank all the contributors, invited speakers and participants for support and co-operation.

Kolkata, India	Syamal Kumar Dana
Kolkata, India	Prodyot Kumar Roy
Potsdam, Germany	Jürgen Kurths

Contents

Part I   Data Analysis

A Unified Approach to Attractor Reconstruction   3
Louis M. Pecora, Linda Moniz, Jonathan Nichols and Thomas L. Carroll

Multifractal Analysis of Physiological Data: A Non-Subjective Approach .   21
G. Ambika, K.P. Harikrishnan and R. Misra

Direction of Information Flow Between Heart Rate, Blood Pressure and Breathing   33
Teodor Buchner, Jan J. Zebrowski, Grzegorz Gielerak and Monika Grzęda

Part II   Cardiovascular Physics: Modelling

The Mathematical Modelling of Inhomogeneities in Ventricular Tissue ...   51
T.K. Shajahan, Sitabhra Sinha and Rahul Pandit

Controlling Spiral Turbulence in Simulated Cardiac Tissue by Low-Amplitude Traveling Wave Stimulation   69
Sitabhra Sinha and S. Sridhar

Suppression of Turbulent Dynamics in Models of Cardiac Tissue by Weak Local Excitations   89
E. Zhuchkova, B. Radnayev, S. Vysotsky and A. Loskutov

Synchronization Phenomena in Networks of Oscillatory and Excitable Luo-Rudy Cells 107
G. V. Osipov, O. I. Kanakov, C.-K. Chan, J. Kurths, S. K. Dana,
L. S. Averyanova and V. S. Petrov

Nonlinear Oscillations in the Conduction System of the Heart – A Model . 127
Krzysztof Grudziński, Jan J. Żebrowski and Rafał Baranowski

Part III   Cardiovascular Physics: Data Analysis

Statistical Physics of Human Heart Rate in Health and Disease 139
Ken Kiyono, Yoshiharu Yamamoto and Zbigniew R. Struzik

**Cardiovascular Dynamics Following Open Heart Surgery: Early
Impairment and Potential for Recovery** 155
Robert Bauernschmitt, Niels Wessel, Hagen Malberg, Gernot Brockmann,
Jürgen Kurths, Georg Bretthauer and Rüdiger Lange

**Application of Empirical Mode Decomposition to Cardiorespiratory
Synchronization** 167
Ming-Chya Wu and Chin-Kun Hu

Part IV   Cognitive and Neurosciences

**Brain Dynamics and Modeling in Epilepsy: Prediction
and Control Studies** 185
Leonidas Iasemidis, Shivkumar Sabesan, Niranjan Chakravarthy,
Awadhesh Prasad and Kostas Tsakalis

An Expressive Body Language Underlies *Drosophila* Courtship Behavior . 215
Ruedi Stoop and Benjamin I. Arthur

Speech Rhythms in Children Learning Two Languages 229
T. Padma Subhadra, Tanusree Das and Nandini Chatterjee Singh

The Role of Dynamical Instabilities and Fluctuations in Hearing 239
J. Balakrishnan

Electrical Noise in Cells, Membranes and Neurons 255
Subhendu Ghosh, Anindita Bhattacharjee, Jyotirmoy Banerjee,
Smarajit Manna, Naveen K. Bhatraju, Mahendra. K. Verma
and Mrinal K. Das

Index 269

Contributors

G. Ambika Indian Institute of Science Education and Research, Pune-411 008, India, g.ambika@iiserpune.ac.in

Benjamin I. Arthur Jr. Physics Department, Institute of Neuroinformatics, University/ETH Zürich, Winterthurerstr. 190, 8057 Zürich, Switzerland, ruedi@ini.phys.ethz.ch

L.S. Averyanova Department of Radiophysics, Nizhny Novgorod University, 23, Gagarin Avenue, 603950 Nizhny Novgorod, Russia, l.averyanova@gmail.com

J. Balakrishnan School of Physics, University of Hyderabad, Central Univeristy P.O., Gachi Bowli, Hyderabad 500 046, India, jbsp@uohyd.ernet.in, janaki05@gmail.com

Jyotirmoy Banerjee Department of Biophysics, University of Delhi South Campus, Benito Juarez Road, New Delhi 110021, India, j_bapi@rediffmail.com

Rafał Baranowski Institute of Cardiology, ul.Alpejska 42, Warszawa, Poland, rbaranowski@ikard.pl

Robert Bauernschmitt Department of Cardiovascular Surgery, German Heart Centre, Lazarettstraße 36, 80636 Munich, Germany, bauernschmitt@dhm.mhn.de

Naveen K. Bhatraju Department of Animal Sciences, School of Life Sciences, University of Hyderabad, Hyderabad 500046, India, naveenb82@gmail.com

Anindita Bhattacharjee Department of Mathematics, Jadavpur University, Kolkata 700032, India, animili@rediffmail.com

Georg Bretthauer Institute for Applied Computer Science, Karlsruhe Research Center, 76021 Karlsruhe, Germany, bretthauer@iai.fzk.de

Gernot Brockmann Department of Cardiovascular Surgery, German Heart Centre, Lazarettstrasse 36, 80636 Munich, Germany, brockmann@dhm.mhn.de

Teodor Buchner Faculty of Physics, Warsaw University of Technology, ul. Koszykowa 75, 00-662 Warsaw, Poland, buchner@if.pw.edu.pl

Thomas L. Carroll Naval Research Laboratory, Washington, DC 20375, USA, Thoms.L.Carroll@anvil.nrl.navy.mil

Niranjan Chakravarthy Department of Electrical Engineering, Arizona State University, Tempe, Arizona, USA, nchakra@asu.edu

C.-K. Chan Institute of physics, Academia Sinica, Nankang, Taipei 11529, Taiwan, ckchan@gate.sinica.edu.tw

Mrinal K. Das Institute of Informatics, University of Delhi South Campus, Benito Juarez Road, New Delhi 110021, India, das_mkd@yahoo.com

Tanusree Das National Brain Research Centre, Near NSG Campus, Nainwal Mode, NH-8, Manesar, 122 050, Haryana, India, tanusree@nbrc.res.in

S.K. Dana Central instrumentation, Indian Institute of Chemical Biology, Jadavpur, Kolkata, 700032, India, skdana@iicb.res.in

Subhendu Ghosh Department of Biophysics, University of Delhi South Campus, Benito Juarez Road, New Delhi 110021, India, profsubhendu@gmail.com

Grzegorz Gielerak Department of Internal Diseases and Cardiology, Military Institute of the Health Services, Szaserów 128, 00-909 Warsaw, Poland, gielerak@wim.mil.pl

Krzysztof Grudziński Physics of Complex Systems, Faculty of Physics, Warsaw University of Technology, ul.Koszykowa 75, Warszawa, Poland, kgrudz@aster.pl

Monika Grzęda Department of Internal Diseases and Cardiology, Military Institute of the Health Services, Szaserów 128, 00-909 Warsaw, Poland, gmonia@o2.pl

K.P. Harikrishnan Department of Physics, The Cochin College, Cochin-682 002, India, kp_hk2002@yahoo.co.in

Leonidas Iasemidis The Harrington Department of Bioengineering and Electrical Engineering, Arizona State University, Tempe, Arizona, USA, Leon.iasemidis@asu.edu

Contributors

Chin-Kun Hu Institute of Physics, Academia Sinica, Nankang, Taipei 11529, Taiwan; Center for Nonlinear and Complex Systems and Department of Physics, Chung-Yuan Christian University, Chungli 32023, Taiwan, huck@phys.sinica.edu.tw

O.I. Kanakov Department of Radiophysics, Nizhny Novgorod University, 23, Gagarin Avenue, 603950 Nizhny Novgorod, Russia, okanakov@rf.unn.ru

Ken Kiyono College of Engineering, Nihon University, 1 Naka-gawara, Tokusada, Tamura-machi, Koriyama City, Fukushima, 963-8642, Japan, kiyono@ge.ce.nihon-u.ac.jp

Jürgen Kurths Department of Physics, University of Potsdam, Am Neuen Palais 10, 14415 Potsdam, Germany, jkurths@agnld.uni-potsdam.de

Rüdiger Lange Department of Cardiovascular Surgery, German Heart Centre, Lazarettstrasse 36, 80636 Munich, Germany, lange@dhm.mhn.de

A. Loskutov Physics Faculty, Moscow State University, Moscow 119992, Russia, loskutov@chaos.phys.msu.ru

Hagen Malberg Institute for Applied Computer Science, Karlsruhe Research Center, 76021 Karlsruhe, Germany, malberg@iai.fzk.de

Smarajit Manna Department of Biophysics, University of Delhi South Campus, Benito Juarez Road, New Delhi 110021, India, s.smarajit@gmail.com

R. Misra Inter University Centre for Astronomy and Astrophysics, Post Bag 4, Ganeshkhind, Pune 411 007, India, rmisra@iucaa.ernet.in

Linda Moniz Department of Mathematics, Trinity College, Washington, DC 20017, USA, MonizL@Trinitydc.edu, Linda.Moniz@jhuapl.edu

Jonathan Nichols Naval Research Laboratory, Washington, DC 20375, USA, jonathan.nichols@nrl.navy.mil, pele@ccs.nrl.navy.mil

G.V. Osipov Department of Radiophysics, Nizhny Novgorod University, 23, Gagarin Avenue, 603950 Nizhny Novgorod, Russia, osipov@vmk.unn.ru

Rahul Pandit Centre for Condensed Matter Theory, Department of Physics, Indian Institute of Science, Bangalore 560012, India, rahul@physics.iisc.ernet.in

Awadhesh Prasad Department of Physics and Astrophysics, University of Delhi, Delhi, India, awadhesh@physics.du.ac.in

Louis M. Pecora Naval Research Laboratory, Washington, DC 20375, USA, pecora@anvil.nrl.navy.mil

V.S. Petrov Department of Radiophysics, Nizhny Novgorod University, 23, Gagarin Avenue, 603950 Nizhny Novgorod, Russia, valentin.s.petrov@gmail.com

B. Radnayev Physics Faculty, Moscow State University, Moscow 119992, Russia, borisrad@yandex.ru

Shivkumar Sabesan Department of Electrical Engineering, Arizona State University, Tempe, Arizona, USA, ssabesa@asu.edu

T.K. Shajahan Centre for Condensed Matter Theory, Department of Physics, Indian Institute of Science, Bangalore 560012, India, shajahan@physics.iisc.ernet.in, shajahan.tk@gmail.com

Nandini Chatterjee Singh National Brain Research Centre, Near NSG Campus, Nainwal Mode, NH - 8, Manesar, 122 050, Haryana, India, nandini@nbrc.ac.in

Sitabhra Sinha The Institute of Mathematical Sciences, CIT Campus, Taramani, Chennai 600113, India, sitabhra@imsc.res.in

S. Sridhar The Institute of Mathematical Sciences, CIT Campus, Taramani, Chennai 600113, India, ssridhar@imsc.res.in

T. Padma Subhadra National Brain Research Centre, Near NSG Campus, Nainwal Mode, NH - 8, Manesar, 122 050, Haryana, India, padma@nbrc.res.in

Ruedi Stoop Institute of Neuroinformatics, Physics Department, University/ETH Zürich, Winterthurerstr. 190, 8057 Zürich, ruedi@ini.phys.ethz.ch

Zbigniew R. Struzik Educational Physiology Laboratory, Graduate School of Education, The University of Tokyo, 7-3-1 Hongo, Bunkyo-ku, Tokyo 113-0033, Japan, z.r.struzik@p.u-tokyo.ac.jp

Kostas Tsakalis Department of Electrical Engineering, Arizona State University, Tempe, Arizona, USA, tsakalis@asu.edu

Mahendra. K. Verma Department of Physics, Indian Institute of Technology, Kanpur 208016, India, mkv@iitk.ac.in

S. Vysotsky Physics Faculty, Moscow State University, Moscow 119992, Russia, saimon@polly.phys.msu.ru

Niels Wessel Department of Physics, University of Potsdam, Am Neuen Palais 10, 14415 Potsdam, Germany, wessel@agnld.uni-potsdam.de

Ming-Chya Wu Research Center for Adaptive Data Analysis, National Central University, Chungli 32001, Taiwan; Institute of Physics, Academia Sinica, Nankang, Taipei 11529, Taiwan, mcwu@ncu.edu.tw

Yoshiharu Yamamoto Educational Physiology Laboratory, Graduate School of Education, The University of Tokyo, 7-3-1 Hongo, Bunkyo-ku, Tokyo 113-0033, Japan, yamamoto@p.u-tokyo.ac.jp

Jan J. Żebrowski Physics of Complex Systems, Faculty of Physics, Warsaw University of Technology, Koszykowa 75, Warszawa, Poland, zebra@if.pw.edu.pl

E. Zhuchkova Physics Faculty, Moscow State University, Moscow 119992, Russia, ekaterina@physik.tu-berlin.de

Part I
Data Analysis

A Unified Approach to Attractor Reconstruction

Louis M. Pecora, Linda Moniz, Jonathan Nichols and Thomas L. Carroll

Abstract In the analysis of complex, nonlinear time series, scientists in a variety of disciplines have relied on a time delayed embedding of their data, i.e. attractor reconstruction. This approach has left several long-standing, but common problems unresolved in which the standard approaches produce inferior results or give no guidance at all. We propose an alternative approach that views the problem of choosing all embedding parameters as being one and the same problem addressable using a single statistical test formulated directly from the reconstruction theorems. This unified approach resolves all the main issues in attractor reconstruction.

Keywords Attractor · time series · embedding · delay coordinates

1 Introduction

One of the most powerful analysis tools for investigating experimentally observed nonlinear systems is attractor reconstruction from time series which has been applied in many fields [1, 2]. The problem of how to connect the phase space or state space vector $\mathbf{x}(t)$ of dynamical variables of the physical system to the time series $s(t)$ measured in experiments was first addressed in 1980 by Packard et al. [3] who showed that it was possible to reconstruct a multidimensional state-space vector by using time delays (or advances which we write as positive delays) with the measured, scalar time series $s(t)$. Thus, a surrogate vector $\mathbf{v}(t) = (s(t), s(t+t), s(t+2t), \ldots)$ for $\mathbf{x}(t)$ could be formed from scalar measurements. This is essentially a particular choice of coordinates in which each component is a time-shifted version of the others with the time shift between adjacent coordinates the same.

Takens [4] and later Sauer et al. [5] independently put this idea on a mathematically sound footing by showing that given any time delay τ and a dimension $\Delta \geq 2$ box-counting dimension $(\mathbf{x}) + 1$ then, nearly all delay reconstructions are one to one and faithful (appropriately diffeomorphic) to the original state space vector $\mathbf{x}(t)$.

L.M. Pecora (✉)
Naval Research Laboratory, Washington, DC 20375
e-mail: pecora@anvil.nrl.navy.mil

These important theorems allow determination of system dynamical and geometric invariants from time series in principle.

The above theorems are existence proofs. They do not directly show how to get a suitable time delay τ or embedding dimension Δ from a finite time series. From the very start [3] emphasis has been put on heuristic reasoning rather than mathematically rigorous criteria for selecting statistics to determine τ and Δ and that remains true up to the present [2]. But many issues are unsettled. For example, there are no clear-cut statistical approaches for dealing with multiple time scales, multivariate data, and avoiding overly long time delays in chaotic data even though these problems are common in many systems and are often acknowledged as important issues in attractor reconstruction.

Various approaches to the τ problem have been autocorrelation, mutual information [6], attractor shape [7, 8], predictive statistics based on various models [9–11], an early attempt to combine finding τ with topological considerations [12], nearest neighbor measures [13], and higher-dimensional version of mutual information [14]. All these approaches have serious short comings (see Kantz and Schreiber [2], Grassberger et al. [14], and Abarbanel [1] for critiques).

To determine Δ Kennel et al. [15, 16] developed false nearest neighbor (FNN) statistics (cf. Ref. [12]). These statistics require one to choose an arbitrary threshold which ignores all structure under that scale which may be considerable for many attractors. Furthermore, in chaotic systems the statistic can be skewed by the divergence of nearby points on the attractor or the existence of two time scales so that the procedure may not truly terminate. Cao [17] has suggested a scale-free FNN approach, but this struggles with properly identifying an asymptote value. And compare Fraser's attempt [18] to go beyond a two-dimensional statistic by using redundancies which has difficulties with computation with probability distributions when the embedding dimension increases.

The problem of a long time window in chaotic systems is often mentioned in passing, but only with simple admonitions that somehow the product $\tau \Delta$ should not get "too large." Such problems point to the fact that until now there have not been statistics to check on overly long embedding times or to guide the user on the quality of the attractor reconstruction.

Very little has been done with multivariate time series. Usually data analysis simply extends what is done for univariate time series [9, 19] thus retaining the shortcomings. There are very few tests for optimal choices of time series to use from a multivariate set (outside of eliminating linear dependence using singular value decomposition [20]) or the Gamma test of Jones [21, 22]).

Finally time series with multiple time scales pose another difficult and unaddressed problem (an exception is the more recent work in Ref. [10], although those approaches rely on specific models).

We claim that the above approaches to attractor reconstruction often artificially divide the problem into two problems (finding find τ and Δ separately) thereby causing more difficulties than necessary while failing to address common problems of multiple time series, overembedding [23, 24] and multivariate data. In this paper we show that only one criterion is necessary for determining embedding parameters. This single criterion is the determination of functional relationships among the components of $\mathbf{v}(t)$ (the reconstruction vector). This allows us to find τ and Δ simultaneously and deal with the unsolved problems of multivariate data, excessively large $\tau\Delta$ and multiple time scales. For this reason we refer to our approach as a unified approach to attractor reconstruction. Our method does not rely on choice of a threshold or some number of nearest neighbors, but rather requires the practitioner to choose a level of confidence for his or her application of the attractor reconstruction. Each level of confidence will be dependent on the application and context and is independent of the method presented here.

This paper is a somewhat condensed version of work that appeared in CHAOS [25].

2 Continuity and Undersampling Statistics

2.1 Continuity or Function Statistic

Takens [4] showed that, generically, $s(t+\tau)$ is functionally independent of $s(t)$. The requirement of any embedding or attractor reconstruction is that one must construct vectors which have independent coordinates. We take this to be the central criterion since without it there is no embedding or reconstruction. This criterion is explicitly stated in Takens theorem [4] (p. 369) and echoed again in the embedding theorems of Sauer et al. [5] (p. 582) which extend the embedding theorems to arbitrary compact invariant sets with mild conditions.

With this in mind we can provide a general requirement which quantifies the functional independence of reconstruction coordinates. Consider a multivariate time series data set $\{s_i(t) | i = 1, \ldots, M\}$ sampled simultaneously at equally-spaced intervals, $t = 1, \ldots, N$. Then for each $\mathbf{v}(t)$ component we can sequentially choose among M time series and various delays (not necessarily equal). Thus, suppose we have a $\mathbf{v}(t)$ of d dimensions $\mathbf{v}(t) = (s_{j_1}(t+\tau_1), s_{j_2}(t+\tau_2), \ldots, s_{j_d}(t+\tau_d))$, where the $j_k \in \{1, \ldots, M\}$ are various choices from the mutivariate data and, in general, each τ_k is different for each component; usually $\tau_1 = 0$. To decide if we need to add another component to $\mathbf{v}(t)$, i.e. increase the dimension Δ of the embedding space, we must test whether the new candidate component, say $s_{j_{d+1}}(t+\tau_{d+1})$ is *functionally independent* of the previous d components. Mathematically we want to test the equality,

$$s_{j_{d+1}}(t+\tau_{d+1}) \underset{?}{=} f(s_{j_1}(t+\tau_1), s_{j_2}(t+\tau_2), \ldots, s_{j_d}(t+\tau_d)), \qquad (1)$$

for any function $f : \mathbf{R}^d \to \mathbf{R}^1$. Equation (1) is the rigorous criterion for generating new $\mathbf{v}(t)$ components and is a general requirement for independent coordinates [26]. Using this approach we continue to add components to $\mathbf{v}(t)$ until all possible candidates for new components from all time series and for all τ values are functions of the previous components. In this way we have found Δ and all delays *simultaneously* and are done. Note, we have not separated the τ and Δ problems – they are found together. Below we develop a statistic to select $s_{j_{d+1}}$ and τ_{d+1} that fulfill the independence criterion.

In the case of embedding of finite data, a distinction must be made between the dimension of the dynamical system that produces the data (and an appropriate embedding dimension for it) and of the reconstructed object that the *data* describes. The aim of the procedure described here is both to make that distinction and to embed the data so that the reconstruction is in the maximum dimension *guaranteed by the data to exhibit independent coordinates*, but no higher or lower. The procedure described herein also excludes the delays that are explicitly forbidden by the embedding theorems, that is, delays equal to the period or twice the period of a periodic orbit.

2.2 Continuity Statistic

The expression of a mathematical property (Eq. (1)) as a statistic is a necessary step if we want to analyze data for such a property. This inevitably involves a heuristic step and the introduction of extra parameters or thresholds which are unavoidable. Our objective is to make the heuristic a strong one in that the statistic should capture the mathematics and provide a confidence level. Thus we aim for a test for functional independence at some maximal or optimal level. We do this in analogy with the linear case (for example, singular value decomposition) in which we can find many linearly independent coordinates, but we choose those whose independence is maximum in some sense.

We build our test statistic for functional relationships (Eq. (1)) on the simple property of continuity. It captures the idea of a function mapping nearby points in the domain to nearby points in the range yet assumes nothing more about the function. We use a version of a continuity statistic [27, 28] with a new null hypothesis [29] that is more flexible in the presence of noise.

For example, suppose $\mathbf{v}(t)$ has d coordinates and we want to test if we need to add another $d + 1$st component. Equation (1) is a test for evidence of a continuous mapping $\mathbf{R}^d \to \mathbf{R}^1$ (Eq. (1)). Following the definition of continuity we choose a positive ε around a fiducial point $s_{j_{d+1}}(t_0 + \tau)$ in $\mathbf{R}^1$ (τ is fixed for now). We pick a δ around the fiducial point $\mathbf{v}(t_0) = (s_{j_1}(t_0 + \tau_1), \ldots, s_{j_d}(t_0 + \tau_d))$ in $\mathbf{R}^d$ corresponding to $s_{j_{d+1}}(t_0 + \tau)$. Suppose there are k points in the domain δ set. Of these suppose l points land in the range ε set. We invoke the continuity null hypothesis that those l points landed in the ε set by chance with probability p. This is shown schematically in Fig. 1. A good choice for data which we use here is $p = 0.5$, i.e. a coin flip on whether the points are mapped from δ into ε. Other choices are possible [27,29], but $p = 0.5$ is actually a standard null, harder to reject and very robust under additive noise. Now we pick the confidence level α at which we reject the null hypothesis. If the probability of getting l or more points in the ε set (a binomial distribution) is

A Unified Approach to Attractor Reconstruction

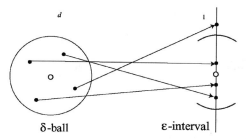

Fig. 1 The δ and ε sets and their data points in R^d and R^1 respectively. Of the k points ($k = 4$, here) in the δ-ball l ($l = 3$, here) are mapped into the ϵ interval and 1 is mapped outside. The probability of interest by which we accept or reject the null hypothesis is the cumulative binomial for getting 3 or more points in the ε interval with probability p (see text)

less than α, we reject the null. For example, if we choose $\alpha = 0.05$, this insures that there are at least 5 points in both δ and ε balls to reject the null hypothesis.

Table 1 shows the number of points in the δ-set that must be in the ε-set to reject the null hypothesis at the $\alpha = 0.05$ level. Note that not requiring all δ-points to map into the ε-set makes the statistic less susceptable to noise or other measurement problems. We comment more on choosing confidence levels below. We note here that an essential part of the reconstruction process is to report not only delays and embedding dimension, but the null hypotheses and the confidence levels used. We do this throughout. We repeat the above process by decreasing ε and varying δ until we cannot reject the null. Call the smallest scale at which we can reject the null our **continuity statistic** ε^*. We sample the data sets at many fiducial points and calculate the average $<\varepsilon^*>$. The data determines ε^*, the smallest scale for which we can claim a function relationship. If the time series is too short to support an embedding, this will be reflected in large and uniform ε^* for any delay.

Table 1 The number of points from the δ-set that must map into the ε-set to reject the null hypothesis at the 95% confidence level ($\alpha = 0.05$). The number of δ points and ε points are the same from 5 to 7, after which the number of δ points can be less than the number of ε points, i.e. one or more δ points can be mapped out of the ε-set and the null hypothesis can still be rejected. Below 5 points the significance is less than 95% and the null hypothesis cannot be rejected

Number of δ points	Number of points in ε-set	Number of points not in ε-set
5	5	0
6	6	0
—	—	—
8	7	1
9	8	1
10	9	1
11	9	2
12	9	3
13	10	3

We suggest that a way to view this statistic is analogous to how we look upon tests for linear independence. If we examine normalized covariances of separate time series, we rarely see either 1's or 0's, but rather numbers between 0 and 1. The closer to 0 a covariance is the less linearly dependent are the two associated time series. In this more general case, the larger $<\varepsilon^*>$ the more functionally independent are two coordinates.

If we succeed in reducing $<\varepsilon^*>$ by adding components at proper delays we will know we are doing well in reconstructing the attractor. When we can no longer reduce $<\varepsilon^*>$, we are done – this is the best we can do with the given data set. Specifically, we apply the continuity test sequentially, we start with one component and build up $\mathbf{v}(t)$ one dimension at a time by examining the continuity statistic $<\varepsilon^*>$ as a function of delay and/or data set. If possible we choose τ at a local maximum of $<\varepsilon^*>$ to assure the most independent coordinates (as in Eq. (1)). If $<\varepsilon^*>$ remains small out to large τ , we need not add more components; we are done and $\Delta = d$. In the multivariate case we can extend this to generating an $<\varepsilon^*>$ for each potential new component $s_j(t + \tau), j = 1, \ldots, M$ of $\mathbf{v}(t)$ and choose that component which has the least functional relationship (a maximum of $<\varepsilon^*>$) to the previous d components.

2.3 Comments on Choosing a Confidence Level

All statistics and tests including those for finding reconstruction parameters will require some threshold or numerical parameters to be chosen. These can be deciding on the number of points to examine locally, setting a FNN threshold, or choosing a method of averaging quantities that mitigates against spurious behavior. However, the amount of ambiguity over these choices can be greatly diminished by choosing a null hypothesis that directly captures the central mathematical property to be tested and then setting a confidence level for rejecting or accepting that null hypothesis. This latter approach is better simply because it can be evaluated in terms of probabilities rather than absolute quantities which give no inkling of confidence level. That is, one has a gauge into how likely the results are to be valid.

In our case the null hypothesis and confidence level α determine the number of local points necessary to reject the null hypothesis and accept that there is a mapping at some ε^* level (see Table 1. above). It is important to report the null hypothesis and the confidence level along with the ε^* values.

Choosing an α value a priori is generally not possible. The usual choices are 95% or 99% confidence levels, $\alpha = 0.05$ and $\alpha = 0.01$, respectively, since these are standards in statistics research as they relate to standard deviations in normal probability distributions. However, in some cases one can do a risk analysis or use decision theory to determine α. That is, one knows the rewards and penalties for choosing correctly or incorrectly regarding the null hypothesis. By choosing a desired reward lower bound or penalty upper bound we can calculate what value of α we should use. Such considerations are common in medical, engineering, and communications decisions and analysis [30]. For more information on decision theory one should consult Jaynes [31].

Our purpose here is not to enter into decision theory since that takes us beyond attractor reconstruction and will be specific to each application and its context. Instead we note that if a risk analysis is available, then it will provide a confidence level that we can immediately use in the determination of best (in the sense of the risk) reconstruction. To put this another way, we push the choice of the "cutoff" parameter α further up to the user of the method since that is where a confidence level can be sensibly chosen. We contend that this is an advantage to our approach along with the use of a statistic close to the original mathematical requirement.

2.4 A Self-Consistent Test, the Undersampling Statistic

In principle the continuity statistic should be the only test we need since it not only determines a good set of $\mathbf{v}(t)$ components, but also a "stopping point" when there are no more independent components, hence automatically giving Δ and τ together. But real data is finite in number and resolution. Because one cannot get nearest neighbors arbitrarily close for finite data, eventually, with large enough delays any two reconstruction vectors from a chaotic time series will necessarily have some components appear randomly scattered on the attractor. This phenomenon is endemic to all chaotic systems. Mathematically the problem is that in looking at points distant in time we are essentially looking at points distributed on a high iteration of the flow or map. This indicates the manifold of the system is very folded and contorted and we are undersampling it. Thus, one cannot discern a loss in functional relation from an undersampled manifold both of which give a large $<\varepsilon^*>$. It is easy to see this effect in a simple one-dimensional map like the logistic map. Kantz and Olbrich [23] showed a similar overembedding phenomenon for simple maps in high dimensions.

We developed an undersampling statistic to detect when $\mathbf{v}(t)$ components enter this undersampling regime. We use the null hypothesis that at least one of the $\mathbf{v}(t)$ components is randomly distributed on the attractor. We test to see if componentwise difference between a point $\mathbf{v}(t_0)$ and its nearest neighbor $\mathbf{v}_{NN}$ in $\mathbf{R}^{d+1}$ (the combined δ and ε spaces) is on the order of typical distances between randomly distributed points on the attractor. To do this test we generate the baseline probability distribution of differences between randomly chosen points from each time series. It is easy to derive the distribution $\sigma_j(\xi)$ of differences ξ between randomly chosen points from the probability distribution $\rho_j(x)$ of values in the jth time series. We calculate $\rho_j(x)$ for each time series $j = 1, \ldots, M$ (e.g. by binning the data). Then it is easy to show under the assumptions of randomly chosen points that $\sigma(\xi) = \int \rho(x)\rho(\xi - x)dx$. Then for each component i of the difference between a point $\mathbf{v}(t_0)$ and its nearest neighbor $\mathbf{v}_{NN} - \mathbf{v}(t_0)$ the probability of getting a value less than or equal to $\xi_i = |(\mathbf{v}_{NN} - \mathbf{v}(t_0))_i|$ (the *NN* distance for the ith component) at random from the j_ith time series is $\Gamma_i = \int_0^{\xi_i} \sigma_{j_i}(\zeta)d\zeta$. Let $\Gamma = \max\{\Gamma_i\}$. As with the continuity statistic we average Γ over fiducial points on the reconstruction.

We choose the confidence level β for rejection or acceptance of the undersampling null hypothesis with the comments of the previous section in mind. We often choose the standard level for rejection at $\beta = 0.05$, so that if $\Gamma \leq 0.05$ we reject the null hypothesis and accept that the time delays are not too large so that the average distances do *not* appear to be distributed in a random fashion. We do this at times in the following examples, but we also display another approach that is an advantage of our unified method. This is to monitor Γ vs. τ as we add reconstruction components (dimensions) and when Γ increases precipitously we stop.

3 Applications

We now present the continuity statistic $<\varepsilon^*>$ and the undersampling statistic Γ for some typical systems of increasing complexity using univariate and multivariate data. Our notation for labeling different $<\varepsilon^*>$ results for Eq. (1) is to simply label the current delays in $\mathbf{v}(t)$. So 0,10 labels the $<\varepsilon^*(\tau)>$ which is testing for a functional relation from the two-dimensional $\mathbf{v}(t)$ (the time series and a 10-step advanced shifted version) to the next possible third-dimensional component time shifted by τ. All time series are normalized to zero mean and standard deviation of 1. For all our statistics our data points are gathered using a temporal exclusion window [32] or using strands in place of points [15] to avoid temporal correlations.

3.1 Quasiperiodic, Multiple Time-Scale System

A non-trivial case is a quasiperiodic system with different time scales. Figure 2 inset shows the torus of this system. The slow and fast times are in the ratio of $2.5\pi:1$ (approximately 8:1) and the time series is sampled at 32 points per fast cycle. The continuity statistic shows only four dimensions are needed since $<\varepsilon^*>$ falls to and remains at a low level after adding the $\tau_4 = 75$ component. This is correct since any 2-Torus can be embedded in four dimensions or less. The two time scales are correctly captured in the ratio of $\tau_2 : \tau_3$. In comparison, using the standard constant delay from the first minimum of the mutual information takes more than 8 dimensions to embed the torus in the sense that both continuity statistics (indicating that independent coordinates can still be added) or false near neighbour statistics (indicating that the torus has been completely unfolded) do not reach their respective terminal values in fewer than 8 dimensions using the standard constant delay.

In Fig. 3, we plot three of the components of the torus attractor using delays of 0, 8, and 67. The attractor fills space equally in all directions and shows only slight self-intersection (hence the need for a fourth dimension with $\tau_4 = 75$). In Fig. 4 the

A Unified Approach to Attractor Reconstruction

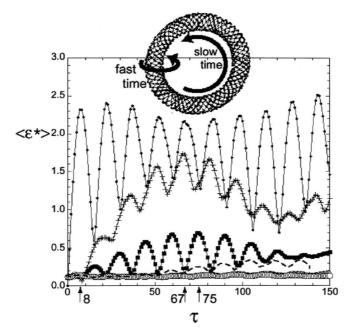

Fig. 2 Continuity statistic for a quasiperiodic system. Inset: The torus for the quasiperiodic system with two time scales. The continuity statistic for the quasiperiodic system for the embeddings 0 (—●—); 0, 8 (—+—); 0, 8, 67 (—■—); 0, 8, 67, 75 (—▽—); and 0, 8, 67, 75, 112 (—○—); and the constant $\tau = 8$ embedding (— — —) using 8 dimensions. The first four delays are noted in the figure

reconstruction uses the standard constant delay of the first minimum of the mutual information resulting in delays of 0, 8, and 16. Note, that in some views (part (b)) the attractor is rather flat and more self intersection appears. This is consistent with the fact that $<\varepsilon^*>$ does not decrease until we have added many more components.

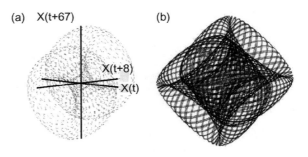

Fig. 3 The attractor reconstruction using delays 0, 8, and 67. Part (**a**) is just points and part (**b**) are connecting trajectories through those points

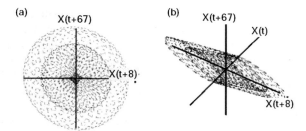

Fig. 4 The attractor reconstruction using delays 0, 8, and 16 as would be done in a constant delay reconstruction

3.2 Lorenz Attractor Reconstruction

We tested the unified approach on a chaotic three-dimensional Lorenz system [33] with $\sigma = 10$, $b = 8/3$, $\rho = 60$. The x-time series was generated using a 4th-order Runge-Kutta solver with a time step of 0.02 and 64,000 points. We calculated $<\varepsilon^*>$ by averaging ε^* over 500 random points on the reconstructions. The results are shown in Fig. 5. This system is chaotic with $<\varepsilon^*>$ eventually increasing with τ because of undersampling so we add the undersampling statistic, Γ, at the bottom of the figure.

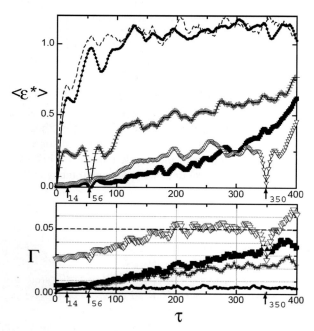

Fig. 5 Continuity $<\varepsilon^*>$ and undersampling Γ statistics for Lorenz time series plus $<\varepsilon^*>$ when 16% gaussian white noise is added to time series (- - - -). Advances: 0 (—●—); 0, 56 (—┼—); 0, 56, 14 (—■—); 0, 56, 14, 350 (—▽—). The delays selected are shown along the τ axes

A three-dimensional $\mathbf{v}(t)$ significantly lowers the $<\varepsilon^*>$ value out to near 300 time steps where it begins to rise. If we try to add another $\mathbf{v}(t)$ component, for example, with a $\tau = 350$, the undersampling statistic rises drastically and goes above the 5% confidence level indicating that the increases in $<\varepsilon^*>$ result from a folded manifold that is undersampled. Nonetheless, the time window over which we have a good embedding is rather wide, about 300 time steps. Using the first minimum in the mutual information ($\tau \approx 16$) would require a $\mathbf{v}(t)$ of about 16 components to accomplish the same reconstruction with constant τ embeddings in the sense that independent coordinates are not achieved with fewer than 16 time delays. In other words, the constant $-\tau$ embedding does not unfold the attractor as efficiently. The continuity statistic provides confidence that details in the attractor down to a scale of 0.2 attractor standard deviation is real. This detail can be lost using arbitrary thresholds of other tests (e.g. mutual information and FNN). The dashed line shows the effect of gaussian white noise of 16% of the time-series standard deviation added to the data. Despite this high noise level much of the $<\varepsilon^*>$ structure remains. This noise robustness comes from our choice of the $p = 0.5$ probability for the $<\varepsilon^*>$ null hypothesis. Similar results occur from shorter time series.

Figure 6 shows the Lorenz attractor from two different views. In Fig. 6(a) we see the "standard" double-winged attractor shape. In Fig. 6(b) we see that the third dimension is not flat, but more extended than plots of the three Lorenz components (x, y, z) because of the use of a longer delay (56). Figure 6(b) displays a third component that is more independent of the first two components.

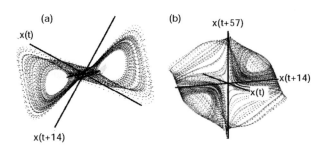

Fig. 6 The Lorenz attractor using delays 14 and 56

3.3 A Multivariate Test Case

For the case of multivariate time series using all three of the Lorenz system's components, x, y, and z we calculated $<\varepsilon^*>$ for all possible combinations of components allowing for time delays or advances (± 100 time steps). The results imply a surprising conclusion. The best set of components for the three-dimensional $\mathbf{v}(t)$ is ($x(0)$, $x(17)$, $z(0)$), i.e. no y component. We emphasize that the "best" here means (in the context of Section 2) the most mutually independent which is our statistical criterion. The $<\varepsilon^*>$ values for the y time series are below those of the x time series at almost all delay values. On reflection we realize that there is no reason why the variables of the physical equations of motion are best for attractor reconstruction.

The numerical tests were done as follows. We started with three cases (x, y, or z) as the first component of the reconstruction vector. We then examined all possible mappings from each of those cases to (x, y, or z) at advanced and delayed times from -100 to $+100$ times steps at 0.02 (in Lorenz system time units) for each step averaging *epsilon** at 400 random points on the attractor for each delay value. After choosing a second component ($x(t + 17)$) we examined all mappings from these two to (x, y, or z) at advanced and delayed times as above. From these we chose $z(t)$. Computationally such a detailed statistical study takes on the order of 10 h on a 1.25 GHz computer. This is certainly not an exhaustive test, since combinatorically more possibilities exist. It is in fact a version of a greedy algorithm [34], but we expect it to be almost optimal because of the low dimension of the Lorenz system.

3.4 An Optimization Approach to the Reconstruction

Thus far we have taken the simple direct route of determining the delays sequentially by choosing relative maxima. This can be characterized as a form of greedy algorithm [34]. A more general and effective way to calculate the proper delays for a time series is to embed the signal in d dimensions, and choose all d delays simultaneously by numerical optimization. In this optimization procedure, the time series is embedded in d dimensions, and the statistic $<\varepsilon^*>$ is calculated for the next $d + 1$st component at each delay from 1 to 100. Since the goal of the unified approach is to minimize the continuity statistic we average the $<\varepsilon^*>$ values for the 100 delays. This delay-averaged quantity $<<\varepsilon^*>>$ can be thought of as the average continuity of the d dimensional embedding to the next ($d + 1$ st) component. This calculation is then done using various d values while monitoring the undersampling statistic.

We chose a downhill simplex method, with simulated annealing [35] (there were many local minima) for the numerical minimization. In order to speed the calculation, rather than vary the radius δ of the d dimensional embedding in $\mathbf{R}^d$, a fixed number of neighboring points on the d dimensional embedding was chosen (40 in this case). From the binomial distribution, if there is a 50%, probability of any particular point landing in the ε set, there is a 5% chance of having 25 points all land in the ε set, so in this routine, ε is the radius on the $d + 1^{st}$ component which contains 25 of the 40 points from the δ neighborhood on the d dimensional embedding.

The minimum value of $<<\varepsilon^*>>$ was calculated from a 30,000 point time series from the Lorenz system, with a time step of 0.03. For the calculation, 10,000 fiducial points were randomly chosen from the 30,000 point time series. For a 3-dimensional embedding, the minimum $<\varepsilon^*>$ was 0.561, at delays of (0, 54, 34). For a 4-dimensional embedding, the minimum $<\varepsilon^*>$ was 0.510, at delays of (0, 52, 35, 68). The optimized approach does show that adding a fourth component is probably warranted although the improvement in the reconstruction is not big. We can compare these with our greedy algorithm given above by multiplying the latter's delays by 2/3 since the Runge-Kutta time step for the latter was 0.02 while here we used 0.03. This gives (integer) delays of (0, 37, 9) in step sizes of 0.03. One pair of delays (37 and 35) agree, but the greedy algorithm resulted in the addition of a short delay whereas the optimized version suggests adding components to the reconstruction vector using larger delays. There were actually many local minima seen using the optimization method, so it is not surprising that different algorithms give different results. There is more than one "good" set of delays for reconstruction, just as there is more than one "bad" set of delays. The optimization approach is more rigorous, but simply choosing local maxima on the $<\varepsilon^*>$ plot works almost as well if there are obvious maxima.

3.5 Neuronal Data

We applied these statistics to neuronal data taken from a lobster's stomatogastric ganglia [36] in Fig. 7. These neurons are part of a central pattern generator and exhibit synchronized bursting similar to Hindmarsh and Rose models [36, 37]. We find that the time delays ($\tau = 0, 278, 110, 214, 15$) range from 15 to 278, a factor of nearly 20. Five dimensions are required for a reconstruction at an undersampling statistic confidence level of $\beta = 0.1$. This agrees qualitatively with a detailed model of the ganglia by Falcke et al. [38]. Using a constant delay established by the 1st minimum of the mutual information ($\tau = 32$) five dimensions still does not capture the long time scales beyond $\tau = 160$, whereas we have structure out to near $\tau = 300$. Adding larger delays (> 300) causes the undersampling statistic to go much higher than 0.1 and we decided to stop with the five delays we have, but we always note what confidence level we used. Figure 8 shows a plot of three of the neuronal reconstruction vector components using 0, 15, and 278. In Fig. 8(a) the spikes are well-resolved into open trajectories that captures their dynamics. From the angle of Fig. 8(a) the long-time undulations appear as a tail because at the time scales of this view they are highly correlated. However, Fig. 8(b) includes a better view capturing the delay of 278. This long delay opens up the tails and exposes the variations of the slow time part of the dynamics. One can see how the spikes "ride" on top of the undulations.

4 Conclusions

Our unified approach uses statistics which faithfully adhere to the rigorous mathematical conditions of the Takens' embedding theorem ($<\varepsilon^*>$) and the accurate geometric picture of excessive time delays (Γ). Both statistics give good indications when we are have a good reconstruction ($<\varepsilon^*>$) and when we have gone beyond what the data will allow (Γ), e.g. undersampling. Hence, unlike other reconstruction approaches we have an indication of the quality of the embedding at a given level of confidence.

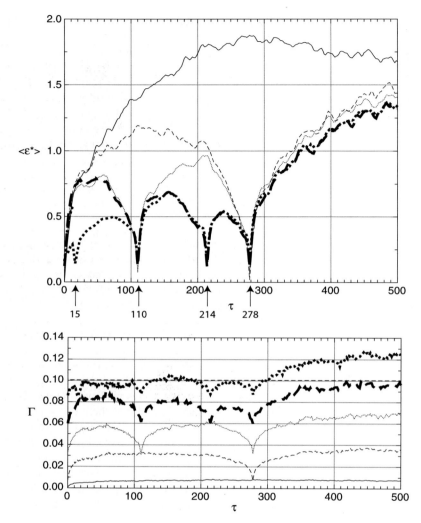

Fig. 7 Continuity $<\varepsilon^*>$ and undersampling Γ statistics for Lobster time series. Advances: 0 (———); 0, 278 (– – – –); 0, 278, 110 (- - - - - -); 0, 278, 110, 214 (_ _ _ _); 0, 278, 110, 214, 15 (■-■-■-■-■). The location of delays 15, 110, 214, and 278 are shown

As we mentioned the $<\varepsilon^*>$ statistic is a generalization of the concept of linear correlation, but note that it is asymmetric in general. For example, given two time series, say $x(t)$ and $y(t)$, we might get very small $<\varepsilon^*>$ for the test of functionality $x(t) \to y(t)$ indicating $y(t)$ is a function of $x(t)$, but get a large $<\varepsilon^*>$ for the reverse relationship $y(t) \to x(t)$ indicating that $x(t)$ is not a function of $y(t)$. A simple example is $y(t) = x^2(t)$.

A Unified Approach to Attractor Reconstruction

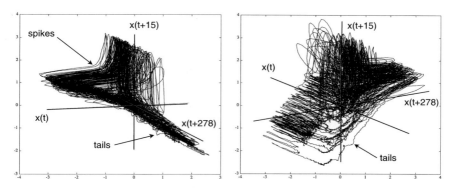

Fig. 8 Two three-dimensional views of the neuronal attractor using delays of 0, 15, and 278. The fast spikes are well resolved and the slow undulations (tails) are open

The statistics $<\varepsilon^*>$ and Γ are data dependent as all statistics should be. Adding more time series points will lower the continuity statistic and improve the undersampling statistic so that the acceptable time window for an embedding will enlarge and more $\mathbf{v}(t)$ components will not be needed. The statistics replace arbitrary time, length, and point-count thresholds with probabilities so that physical scales (which are often unknown) are derived and data-dependent, but not assumed at the outset.

The data-dependent method has the advantage (over current methods) of yielding unambiguous information about the ability to embed a data set at all. If it is not possible to choose successive delays so that the continuity statistics fall, the data set does not include enough information to produce independent embedding coordinates.

We have chosen "standard" statistical confidence levels here of $\alpha = 0.05$ and $\beta = 0.05$ (2 sigmas) in some cases. In others we stop the reconstruction when the undersampling statistic falls precipitously and give the reconstruction parameters along with the level for null hypothesis rejection. In all problems with statistics choosing the level for the null hypothesis rejection is a separate problem and usually depends on each particular situation, hence the confidence level cannot be given a priori, but should be chosen by the end user. Finding the best confidence level would involve doing something like a risk or decision analysis for which payoffs and penalties for correctly or incorrectly rejecting the null hypothesis are known. Whether such an analysis is or is not possible (and often it is not), we believe that these levels should always be give so others can judge the quality of the reconstruction. Because our statistics do not involve arbitrary scales, but rather null hypothesis we can always given the probability that the reconstruction is right.

Finally, we note that the problem of observability for delay embeddings might be addressed with our statistic since in this case there is no closed form solution for chaotic systems [39].

Acknowledgments We would like to acknowledge helpful conversations with M. Kennel, L. Tsimring, H.D.I. Abarbanel, and M.A. Harrison.

References

1. H.D.I. Abarbanel, *Analysis of Observed Chaotic Data*. (Springer, New York, 1996).
2. H. Kantz and T. Schreiber, *Nonlinear Time Series Analysis*. (Cambridge University Press, Cambridge, UK, 1997).
3. N. H. Packard, J.P. Crutchfield, J.D. Farmer et al., *Geometry from a time series*, Physical Review Letters, **45**, 712 (1980).
4. F. Takens, *Detecting Strange Attractors in Turbulence* in *Dynamical Systems and Turbulence, Warwick 1980*, edited by D. Rand and L.-S. Young (Springer, Berlin, 1981), p. 366.
5. T. Sauer, J.A. Yorke, and M. Casdagli, *Embedology*, Journal Statistical Physics, **64**, 579 (1991).
6. A.M. Fraser and H.L. Swinney, *Independent coordinates for strange attractors from mutual information*, Physical Review, **A 33**, 1134 (1986).
7. G. Kember and A.C. Fowler, *A correlation function for choosing time delays in phase portrait reconstructions*, Physica, **D 58**, 127 (1993).
8. M.T. Rosenstein, J.J. Collins, and C.J. De Luca, *Reconstruction expansion as a geometry-based framework for choosing proper delay times*, Physica, **D 73**, 82 (1994).
9. L. Cao, A. Mees, and J.K. Judd, *Dynamics from a multivariate time series*, Physicsa, **D 121**, 75 (1998).
10. K. Judd and A. Mees, *Embedding as a modeling problem*, Physica, **D 120**, 273 (1998).
11. D. Kugiumtzis, *State space reconstruction parameters in the analysis of chaotic time series – the role of the time window length*, Physica, **D 95**(1), 13 (1996).
12. W. Leibert, K. Pawlezik, and H.G. Schuster, *Optimal embeddings of chaotic attractors from topological considerations*, Europhysics Letters, **14**(6), 521 (1991).
13. S.P. Garcia and J.S. Almeida, *Nearest neighbor embedding with different time delays*, Physical Review, **E 71**, 037204 (2005).
14. P. Grassberger, T. Schreiber, and C. Schaffrath, *Nonlinear time sequence analysis*, International Journal of Bifurcation and Chaos in Applied Sciences and Engineering, **1**(3), 521 (1991).
15. M. Kennel and H.D.I. Abarbanel, *False neighbors and false strands: A eliable minimum embedding dimension algorithm*, Physical Review, **E66**, 026209 (2002).
16. M.B. Kennel, R. Brown, and H.D.I. Abarbanel, *Determining embedding dimension for phase space reconstruction using the method of false nearest neighbors*, Physical Review, **A 45**, 3403 (1992).
17. L. Cao, *Practical method for determining the minimum embedding dimension of a scalar time series*, Physica, **D 110**, 43 (1997).
18. A. Fraser, *Information and entropy in strange attractors*, IEEE Transactions on Information Theory, **35**(2), 245 (1989).
19. S. Boccaletti, L.M. Pecora, and A. Pelaez, *Unifying framework for synchronization of coupled dynamical systems*, Physical Review, **E 63**(6), 066219/1 (2001).
20. D. Broomhead and G.P. King, *Extracting qualitative dynamics from experimental data*, Physica, **20 D**, 217 (1986).
21. A.P.M. Tsui, A.J. Jones, and A.G. de Oliveira, *The construction of smooth models using irregular embeddings determined by a gamma test analysis*, Neural Computing & Applications, **10**(4), 318 (2002).
22. D. Evans and A.J. Jones, *A proof of the gamma test*, Proceedings of the Royal Society of London, Series A (Mathematical, Physical and Engineering Sciences), **458**(2027), 2759 (2002).
23. H. Kantz and E. Olbrich, *Scalar observations from a class of high-dimensional chaotic systems: Limitations of the time delay embedding*, Chaos, **7**(3), 423 (1997).
24. E. Olbrich and H. Kantz, *Inferring chaotic dynamics from time-series: On which length scale determinism becomes visible*, Physics Letters, **A 232**, 63 (1997).
25. L.M. Pecora, L. Moniz, J. Nichols et al., *A unified approach to attractor reconstruction*, Chaos, **17**, 013110 (2007).

26. Wendell Fleming, *Functions of Several Variables*. (Springer-Verlag, New York, 1977).
27. L. Pecora, T. Carroll, and J. Heagy, *Statistics for mathematical properties of maps between time-series embeddings*, Physical Review E, **52**(4), 3420 (1995).
28. Louis M. Pecora, Thomas L. Carroll, and James F. Heagy, *Statistics for Continuity and Differentiability: An Application to Attractor Reconstruction from Time Series* in Nonlinear Dynamics and Time Series: Building a Bridge Between the Natural and Statistical Sciences, Fields Institute Communications, edited by C.D. Cutler and D.T. Kaplan (American Mathematical Society, Providence, Rhode Island, 1996), Vol. 11, p. 49.
29. L. Moniz, L. Pecora, J. Nichols et al., *Dynamical assessment of structural damage using the continuity statistic*, Structural Health Monitoring, **3**(3), 199 (2003).
30. D. Middleton, *An Introduction to Statistical Communication Theory*. (IEEE Press, Piscataway, NJ, 1996).
31. E.T. Jaynes, *Probability Theory, The Logic of Science*. (Cambridge University Press, Cambridge, UK, 2003).
32. J. Theiler, *Spurious dimension from correlation algorithms applied to limited time series data*, Physical Review, **A 34**, 2427 (1986).
33. E.N. Lorenz, *Deterministic non-periodic flow*, Journal of Atmospheric Science, **20**, 130 (1963).
34. G. Strang, *Introduction to Applied Mathematics*. (Wellesley-Cambridge Press, Wellesley, MA, 1986).
35. W.H. Press, B.P. Flannery, S.A. Teukolsky et al., *Numerical Recipes*. (Cambridge University Press, New York, 1990).
36. R.C. Elson, A.I. Selverston, R. Huerta et al., *Synchronous behavior of two coupled biological neurons*, Physical Review Letters, **81**(25), 5692 (1998).
37. J.L. Hindmarsh and R.M. Rose, A model of neuronal bursting using three coupled first order differential equations, Proceedings of the Royal Society of London, **B221**, 87 (1984).
38. M. Falcke, R. Huerta, M.I. Rabinovich et al., *Modeling observed chaotic oscillations in bursting neurons: The role of calcium dynamics and IP3*, Biological Cybernetics, **82**, 517 (2000).
39. L. Aguirre and C. Letellier, *Observability of multivariate differential embeddings*, Journal of Physics, **A 38**, 6311 (2005).

Multifractal Analysis of Physiological Data: A Non-Subjective Approach

G. Ambika, K.P. Harikrishnan and R. Misra

Abstract We have recently proposed an algorithmic scheme [22] for the non-subjective computation of correlation dimension from time series data. Here it is extended for the computation of generalized dimensions and the multifractal spectrum and applied to a number of EEG and ECG data sets from normal as well as certain pathological states of the brain and the cardiac system. Comparisons are drawn using a standard low dimensional chaotic system. Our method has the advantage that the analysis is done under identical prescriptions built into the algorithm and hence the comparison of resulting indices becomes non-subjective. This also enables a quantitative characterization of the relative complexity between practical time series such as, those corresponding to the changes in the physiological states of the same system, from the view point of underlying dynamics.

Keywords Time series analysis · generalized dimension · multifractal spectrum

1 Introduction

Over the years, interest has risen in applying the methods and concepts from nonlinear dynamics to problems in physiology. Several topical issues on nonlinear dynamics and physiology are an evidence for this [1–3]. It has opened up new avenues to use methods from nonlinear dynamics as diagnostic tools for the analysis of physiological data. Moreover, physicists and mathematicians have been inspired by the wealth of interesting physiological problems to develop new tools and methods.

Out of the large number of studies done on physiological data, the focus has mainly been on the analysis of EEG and ECG time series data, with the purpose of characterization and prediction from a dynamical systems point of view. The analysis of EEG data from healthy persons and epileptic patients has lead to a better understanding of various aspects of epileptic seizure activities and the corresponding brain states [4, 5], but the question of whether the seizure can be predicted in

G. Ambika (✉)
Indian Institute of Science Education and Research, Pune-411 008, India
e-mail: g.ambika@iiserpune.ac.in

advance is still an open one [6]. There have been a multitude of studies on ECG data sets recorded from healthy persons as well as during some pathological cases, such as, congestive heart disorders and ventricular fibrillation [7–9]. Though most of these studies have indicated evidence for deterministic nonlinearity in the time series for all normal and pathological states of the cardiac system [10, 11], the reliability of these results have also been questioned [12–14] due to various reasons, such as, insufficient data, presence of noise, the subjective nature of the computational techniques and so on.

Various complexity measures have been used to characterize the physiological data. Among them are measures based on recurrence plots [15], symbolic dynamics [16], Lyapunov exponents [17] and correlation dimension [6, 11]. Recently, a new measure based on multiscale entropy has also been introduced [18, 19]. All these methods have their own merits and shortcomings and have been successful in analyzing particular data sets. The most widely used method among these is the one based on the correlation dimension $D(2)$ and have been applied to all types of experimental time series data. In fact, it has remained the most important test statistic for hypothesis testing. But a major drawback with $D(2)$ is the subjective nature of computing $D(2)$ from a time series, as discussed by many authors [20, 21]. To overcome this, we have recently proposed and implemented an algorithmic scheme for the non-subjective computation of $D(2)$ from a time series [22]. Here we extend this scheme for the computation of generalized dimensions $D(q)$ and the $f(\alpha)$ spectrum of a chaotic attractor from a time series. The scheme is first tested using a standard low dimensional chaotic system. It is then applied to an ensemble of EEG and ECG time series from healthy people as well as from those having epilepsy and congestive heart disorders. Our method has the advantage that a non-subjective comparison of the relative complexities in the time series from different states of the brain and the heart is possible. We find that while the normal EEG time series is consistent with pure random noise, there are strong indications of deterministic nonlinearity in the normal heart behavior which shows multifractal character analogous to a low dimensional chaotic system.

The paper is organized as follows: The details of our non-subjective scheme for computing $D(q)$ and the $f(\alpha)$ spectrum from a time series are presented in Section 2 and it is applied on a standard low dimensional chaotic system. The EEG-ECG data sets are then analyzed in detail using the scheme in Section 3. Conclusions are drawn in Section 4.

2 Computation of $D(q)$ and $f(\alpha)$ from Time Series

Our method is based on the Grassberger-Procaccia algorithm for the computation of $D(2)$ from a time series [23, 24]. Here an artificial space of dimension M is constructed from a scalar time series $s(t_i)$ as

$$\vec{x}_i = [s(t_i), s(t_i + \tau), \ldots\ldots\ldots\ldots s(t_i + (M-1)\tau)] \tag{1}$$

where τ is a suitably chosen time delay. The correlation sum is the relative number of points within a distance R from a particular (ith) data point,

$$p_i = \lim_{N_v \to \infty} \frac{1}{N_v} \sum_{j=1, j \neq i} H(R - |\vec{x}_i - \vec{x}_j|) \qquad (2)$$

where N_v is the total number of reconstructed vectors and H is the Heaviside step function. Averaging this quantity over N_c randomly selected $\vec{x}_i$ or centres gives the correlation function

$$C_M(R) = \frac{1}{N_c} \sum_i^{N_c} p(R) \qquad (3)$$

The correlation dimension $D(2)$ is then defined to be

$$D(2) = \lim_{R \to 0} d(\log C_M(R))/d(\log(R)) \qquad (4)$$

which is the scaling index of the variation of $C_M(R)$ with R as $R \to 0$.

To compute $D(2)$, one has to identify a linear part in the $\log C_M(R)$ versus $\log(R)$ plot for each M, called the *scaling region*, which is usually done by the visual inspection of the correlation sum. But in our computational scheme, this is avoided and instead, the scaling region is fixed algorithmically. For this, a maximum value of R, R_{max} and a minimum value of R, R_{min} computed for each M using some criteria based on the algorithm itself and the region between them is taken as the scaling region. There often exists a critical embedding dimension M_{cr} for which $R_{min} \approx R_{max}$, so that significant results can be obtained only for $M \leq M_{cr}$. Hence the computations are done for each value of M starting from $M = 1$ to $M = M_{cr}$. More details regarding the scheme, such as, the criteria for fixing R_{max} and R_{min}, the calculations of error bar etc., can be found elsewhere [22].

As an example, the scheme is first applied to a standard low dimensional chaotic system (Rossler system) modelled by the following set of coupled nonlinear equations:

$$\begin{aligned} \dot{X} &= -Y - Z \\ \dot{Y} &= X + aY \\ \dot{Z} &= b + Z(X - c) \end{aligned} \qquad (5)$$

where a, b and c are the control parameters whose values determine the asymptotic behavior of the system. For $a = b = 0.2$ and $c = 7.8$, the system is known to have an underlying chaotic attractor (Rössler attractor). The result of applying our scheme to the time series consisting of 10,000 data points from the Rössler attractor is shown in Fig. 1.

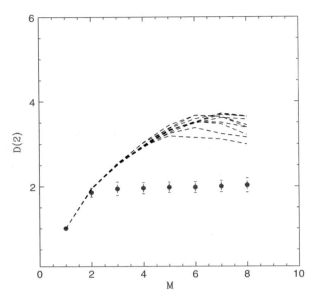

Fig. 1 Results of applying our algorithmic scheme to the time series from Rossler attractor (*filled circle* with error bar) and ten surrogates (*dashed lines*), using 10,000 data points. As expected, the saturated values of $D(2)$ for the data and the surrogates are widely separated

One distinctive advantage of our scheme is that it is more suitable for hypothesis testing as $D(2)$ is computed algorithmically with the same conditions on the data and the surrogates. The result of surrogate analysis using the scheme is also shown in Fig. 1, with ten surrogates for the data. The surrogates were generated applying the Iterated Amplitude Adjusted Fourier Transform (IAAFT) algorithm [25] using the details given in Hegger et al. [26].

A chaotic attractor is characterized by a spectrum of generalized dimensions $D(q)$, where the index q can vary from $-\infty$ to $+\infty$ [27]. $D(q)$ can be converted into a spectrum of singularities called the $f(\alpha)$ spectrum [28], characterizing the global scaling properties of the attractor. Here the parameter α is a continuous variable representing the local scaling index of the fractal set. The transformation from $D(q)$ to $f(\alpha)$ can be shown to be a Legendre transformation (for details, see [27,28]), with the following exact relations: $f_{\max}(\alpha) = D(0)$, $\alpha_{\min} \equiv D(\infty)$ and $\alpha_{\max} \equiv D(-\infty)$. The range of scaling indices ($\alpha_{\max}, \alpha_{\min}$) is an indication of the degree of inhomogeneity present in the attractor.

We now extend our scheme to compute $D(q)$ and the $f(\alpha)$ spectrum from the time series. In order to compute $D(q)$, the correlation sum is first generalized to get

$$C_M^q(R) = \left[\frac{1}{N_c} \sum_i^{N_c} \left(\frac{1}{N_v} \sum_{j=1, j \neq i}^{N_v} H(R - |\vec{x}_i - \vec{x}_j|) \right)^{q-1} \right]^{1/(q-1)} \quad (6)$$

Then the spectrum of dimensions are given by

$$D(q) = \lim_{R \to 0} d(\log C_M^q(R))/d(\log(R)) \tag{7}$$

The average value of $D(q)$ is calculated from the scaling region by taking different values of R. The error in $D(q)$ is also calculated as the mean standard deviation over the average value.

The $D(q)$ values can be calculated for each embedding dimension from $M = 1$ to a critical value M_{cr} using the scheme. For a chaotic attractor, the $D(q)$ values are found to saturate for $M \geq M_{sat}$. To compute the $f(\alpha)$ spectrum, the values of $D(q)$ corresponding to any $M \geq M_{sat}$ can be used.

But since the error also increases with M, we choose $M \geq M_{sat}$ to compute $f(\alpha)$, using the standard equations [27, 28]:

$$\alpha(q) = \frac{d}{dq}[(q-1)D(q)] \tag{8}$$

$$f(\alpha) = q\alpha(q) - (q-1)D(q) \tag{9}$$

In order to get the complete $f(\alpha)$ spectrum, we try to fit the $D(q)$ versus q curve with a continuous and diferentiable function $F(q)$ which is symmetric with respect to some point $q = q0$ and saturates as $q \to \pm\infty$. It is found that such a functional fit can be written in the form:

$$F(q) = AS(q - q_0, \beta) + B \tag{10}$$

with

$$S(q - q_0, \beta) = \frac{(q - q_0)}{|q - q_0|}[e^{-\beta|q-q_0|} - 1] \tag{11}$$

It makes use of four parameters A, B, $q0$ and β which have to be chosen by trial and error depending on the data set to get the best fit curve. Here A and B represent the saturated values of $D(q)$ as $q \to \pm\infty$, q_0 the symmetric point and β represents the slope of the curve at $q = q_0$. A χ^2 fitting is undertaken by adjusting the four parameters until the best fit is obtained indicated by the minimum χ^2 value. The $f(\alpha)$ spectrum is now calculated from this fitted function making use of Eqs. (11) and (12).

To illustrate the method, it is used to compute the $D(q)$ and the $f(\alpha)$ spectrum from the time series of the Rossler attractor. The $D(q)$ values are computed taking a step width of $\Delta q = 0.1$ and is shown in Fig. 2 along with the functional fit for q in the range $(-40, 40)$. The corresponding $f(\alpha)$ spectrum is shown in Fig. 3, which gives $f(\alpha) \equiv D(0) = 2.28$, $\alpha_{min} \equiv D(\infty) = 1.34$ and $\alpha_{max} \equiv D(-\infty) = 3.46$.

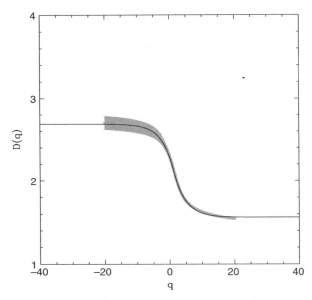

Fig. 2 The generalised dimensions $D(q)$ as a function of q for the time series from the Rössler attractor, along with the functional fit (see text) for $D(q)$. Calculations are done with a step size $\Delta q = 0.1$. The error in the calculation of $D(q)$ is also shown

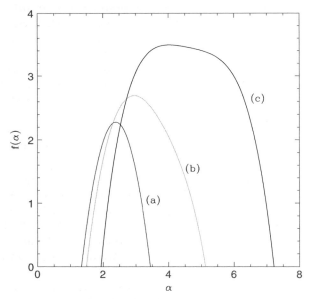

Fig. 3 The $f(\alpha)$ spectrum for the Rossler attractor shown as curve (*a*), computed from the best t curve for $D(q)$ given in Fig. 2. Also shown are the $f(\alpha)$ spectrum of the Rossler time series added with (*b*) 20% white noise and (*c*) 50% white noise. It is clear that, as the percentage of noise increases, the spectrum gets stretched out increasing the range of α values, $(\alpha_{min}, \alpha_{max})$

Experimental time series are usually contaminated with noise and it is important to know the effect of noise on the $f(\alpha)$ spectrum of a chaotic attractor for a proper comparison. To study this, we generate a few data sets by adding different amounts of white noise to the Rossler attractor data. The $f(\alpha)$ spectrum for two of them are shown in Fig. 3 with 20 and 50% addition of white noise respectively. It is clear that as the percentage of white noise increases, the range of α values also increase, making the spectrum wider.

3 Analysis of EEG and ECG Time Series

In this section, we apply this algorithmic scheme to a number of EEG and ECG data sets. The EEG data sets have been downloaded from the website of the Department of Epileptology, University of Bonn while the ECG was obtained from http://www.physionet.org/physiobank/archives. The interest has mainly been on the analysis of EEG data from epileptic patients, as they have shown strong indications of nonlinear behavior.

Here we analyze 3 separate EEG data sets corresponding to different brain states, namely, (a) from normal healthy human beings (b) from epileptic patients during seizure free intervals and (c) from epileptic patients during seizure activity. Each of them consists of continuous data streams of about 24 s long and consisting of approximately 5,000 data points. By performing surrogate analysis using our scheme, we find that the data set (a) from healthy persons coincide with pure random noise and the set (b) from seizure free intervals show deviation from randomness, but do not show saturation for $D(2)$. But the data set (c) during seizure period shows clear indications of nonlinearity with a well saturated value of $D(2)$. The variation of $D(2)$ as a function of M for the seizure case and the surrogates from this set are shown in Fig. 4. For this data set, the $D(q)$ values and the corresponding $f(\alpha)$ spectrum are computed using our scheme. The $D(q)$ values for $M = 3$ are shown in Fig. 5 along with the functional fit. The corresponding $f(\alpha)$ spectrum is shown in Fig. 6. The range of α values are given by $\alpha_{\min} = 0.67$, $\alpha_{\max} = 5.18$ with $D(0) = 2.43$. Also shown in Fig. 6 is the $f(\alpha)$ spectrum for the data corresponding to seizure free intervals computed for $M = 5$.

A comparison with the corresponding indices for the standard low dimensional chaotic attractor shown in Fig. 3 gives some interesting results. During epileptic seizure, the EEG data shows strong deterministic nonlinear behavior with an underlying multifractal attractor. Stochastic fluctuations become dominant for the seizure free intervals though the time series multifractal character. It seems that as the brain tends to be more normal, the percentage of random fluctuations also increase. The time series becomes completely random in the case of healthy people.

We analyse ECG data sets corresponding to two different states of the cardiac system. One set of data was recorded from a healthy heart while the other was from a heart with congestive disorder. Each data set consists of continuous data streams of 5,400 data points with a sampling time of 0.04 s. Results of applying surrogate

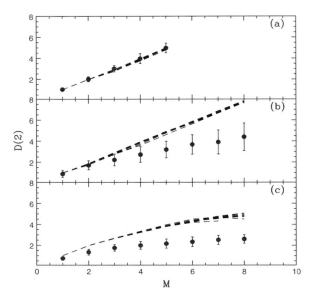

Fig. 4 $D(2)$ as a function of M (*filled circles* with error bar) for the EEG signal for (a) normal brain (b) during seizure free interval (c) during seizure along with the $D(2)$ values for 10 surrogates (*dashed lines*). The total number of points in the data is close to 5,000

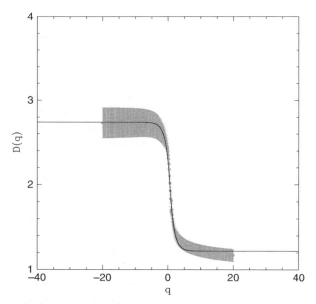

Fig. 5 $D(q)$ as a function of q for the EEG data which show indications of low dimensional chaos, along with the best fit curve for $D(q)$. Note that the error bar in $D(q)$ bulges as $q \to \pm 20$

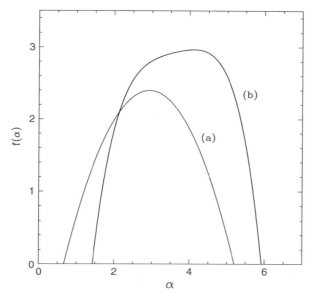

Fig. 6 The $f(\alpha)$ spectrum for the EEG data is shown in (*a*), computed from the best *t* curve for $D(q)$ given in Fig. 5 and in (*b*), the $f(\alpha)$ curve for the EEG data shown in Fig. 4b for the seizure free interval

analysis on the two time series are shown in Fig. 7. It turns out that neither of them are consistent with random behaviour. The saturated value of $D(2)$ for the pathological case and the normal case are found to be 2.04 ± 0.18 and 2.28 ± 0.26. The $f(\alpha)$ spectrum is computed in both cases and the results are shown in Fig. 8. Both data show multifractal character with the range of spectral indices $(\alpha_{\min}, \alpha_{\max})$ given by (0.72, 3.51) and (0.96, 4.98) respectively. Thus, unlike the case of EEG, there is not much difference in the computed $D(2)$ values for data from a normal cardiac system and that from a pathological state, but the range of α values decrease for the latter.

It is generally believed that a normal and healthy physiological system should be more adaptive to the environment and needs to exhibit processes which run of several different time scales to be able to react to an ever changing input [2,9]. In this sense, complexity is a sign of health and hence fluctuations of ageing and pathological systems are likely to show a lesser degree of complexity compared with healthy systems. Thus it is not surprising that the time series from a normal brain, which is the most complex biological system involving spatial and temporal correlations from different parts, is dominated by random fluctuations. The difference in the case of a normal cardiac system may be that its basic function is coupled to the control mechanisms of the body. It involves nonlinear oscillations which swamp out small scale fluctuations present, as indicated by the time series. Also, the appearance of periodicities in the data is associated with the onset of heart disorder [9].

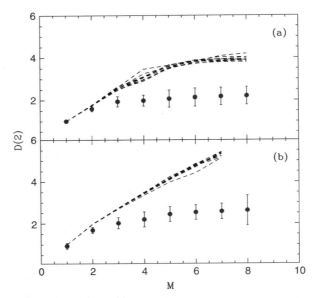

Fig. 7 The $D(2)$ versus M for the two ECG data (a) for healthy heart (b) with longestive disorder along with that for 10 surrogates (*dashed lines*). The total number of data points are 5,400 in both cases

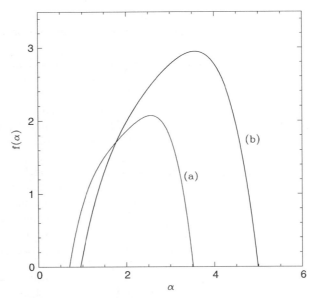

Fig. 8 The $f(\alpha)$ spectrum for the two ECG signals (*a*) for the pathological case and (*b*) for the normal case

4 Conclusions

The main goal of this paper is to present a scheme for the non-subjective computation of $D(q)$ and $f(\alpha)$ spectrum from a time series. It is first applied to a standard low dimensional chaotic system. The change in the $f(\alpha)$ spectrum by the addition of different amounts of white noise is also shown explicitly. The scheme is then used for the analysis of EEG and ECG data sets from healthy people as well as from those having epileptic and cardiac disorders. We find that the EEG data during epileptic seizure shows deterministic nonlinearity and multifractal character with a well defined $f(\alpha)$ spectrum, while that from a normal person is consistent with random noise. On the other hand, the ECG data sets from normal and pathological cases show nonlinearity and multifractal nature, with the spectrum being narrower in the latter case.

As is well known from the multitude of studies in the last decade, it is difficult to give a conclusive evidence for deterministic chaos in experimental systems in general and biological systems in particular. Reasons are many ranging from data acquisition aspects [2] to interpretation of results [19], and chaos can at best be just one candidate among several alternatives even after the rejection of null hypothesis [21, 26]. Thus our scheme can be better utilised as a measure of complexity rather than a diagnostic tool.

An advantage of our scheme is that, since the analysis is done under identical conditions as prescribed by the algorithm, the resulting indices give a better representation for comparison between data sets. This is especially important in the case of physiological data where the changes in the same system, such as for example, due to epileptic or cardiac disorders can be compared in a non-subjective manner.

Acknowledgments KPH and GA acknowledge the hospitality and computing facilities in IUCAA, Pune.

References

1. *Spotlight issue on chaos in the cardiovascular system*, Cardiovasc. Res., **31** (1996).
2. *Special issue on the application of nonlinear dynamics to physiology*, CHAOS, **8** (1998).
3. *Focal issue on mapping and control of complex cardiac arrhythmias*, CHAOS, **12** (2002).
4. N. Pradhan and D. Narayana Dutt, *A nonlinear perspective in understanding the neurodynamics of EEG*, Comput. Biol. Med., **23**, 425 (1993).
5. W. Latzenberger, H. Preissl and F. Pulvermiller, *Fractal dimension of electroencephalographic time series and underlying brain processes*, Biol. Cybernet., **73**, 477 (1995).
6. M.A.F. Harrison, I. Osorio, M.G. Frei, S. Asuri and Y.-C. Lai, *Correlation dimension and integral do not predict epileptic seizures*, CHAOS, **15**, 033106 (2005).
7. L. Glass, A.L. Goldberger, M. Courtemanche and A. Shrier, *Nonlinear dynamics, chaos and complex cardiac arrhythmias*, Proc. R. Soc. Lond. Ser. A, **413**, 9 (1987).
8. F.X. Witkowski, K.M. Karnagh, P.A. Penkoske, R. Plonsey, M.L. Spano, W.L. Ditto and D.T. Kaplan, *Evidence for determinism in ventricular fibrillation*, Phys. Rev. Lett. **75**, 1230 (1995).
9. C.S. Poon and C.K. Merril, *Decrease of cardiac chaos in congestive heart failure*, Nature, **389**, 492 (1997).

10. F. Ravelli and R. Antolini, *Complex dynamics underlying the human electrocardiogram*, Biol. Cybernet., **67**, 57 (1992).
11. R.B. Govindan, K. Narayanan and M.S. Gopinathan, *On the evidence of deterministic chaos in ECG: Surrogate and predictability analysis*, CHAOS, **8**, 495 (1998).
12. L. Glass and P. Hunter, *There is a theory of heart*, Physica D, **43**, 1 (1990).
13. A.L. Goldberger, *Is the normal heart beat chaotic or homeostatic?*, News. Physiol. Sci., **6**, 87 (1991).
14. F. Mitschke and M. Dimming, *Chaos versus noise in experimental data*, Int. J. Bif. Chaos, **3**, 693 (1993).
15. N. Marwan, N. Wessel, U. Meyerfeldt, A. Schirdewan and J. Kurths, *Recurrence plot based measures of complexity and their application to heart rate variability data*, Phys. Rev. E, **66**, 026702 (2002).
16. N. Wessel, C. Ziehmann, J. Kurths, U. Meyerfeldt, A. Schirdewan and A. Voss, *Short term forecasting of life-threatening cardiac arrhythmias based on symbolic dynamics and finite time growth rates*, Phys. Rev. E, **61**, 733 (2000).
17. R. Brown and P. Bryant, *Computing Lyapunov spectrum of a dynamical system from an observed time series*, Phys. Rev. A, **43**, 2787 (1991).
18. M. Costa, A.L. Goldberger and C.K. Peng, *Multiscale entropy analysis of complex physiological time series*, Phys. Rev. Lett., **89**, 068102 (2002).
19. R.A. Thuraisingham and G.A. Gottwald, *On multiscale entropy analysis for physiological data*, Physica A, **366**, 323 (2006).
20. M.B. Kennel and S. Isabelle, *Method to distinguish possible chaos from colored noise and to determine embedding parameters*, Phys. Rev. A, **46**, 3111 (1992).
21. M. Small and K. Judd, *Detecting nonlinearity in experimental data*, Int. J. Bif. Chaos, **8**, 1231 (1998).
22. K.P. Harikrishnan, R. Misra, G. Ambika and A.K. Kembhavi, *A nonsubjective approach to the GP algorithm for analysing noisy time series*, Physica D, **215**, 137 (2006).
23. P. Grassberger and I. Procaccia, *Characterisation of strange attractors*, Phys. Rev. Lett., **50**, 346 (1983).
24. P. Grassberger and I. Procaccia, *Measuring the strangeness of strange attractors*, Physica D, **9**, 189 (1983).
25. T. Schreiber and A. Schmitz, *Improved surrogate data for nonlinearity tests*, Phys. Rev. Lett., **77**, 635 (1996).
26. R. Hegger, H. Kantz and T. Schreiber, *Practical implementation of Nonlinear time series methods: The TISEAN package*, CHAOS, **9**, 413 (1999).
27. T.C. Halsey, M.H. Jensen, L.P. Kadanof, I. Proccacia and B.I. Shraiman, *Fractal measures and their singularities: The characterization of strange sets*, Phys. Rev. A, **33**, 1141 (1986).
28. H. Atmanspacher, H. Scheingraber and N. Voges, *Global scaling properties of a chaotic attractor reconstructed from experimental data*, Phys. Rev. A, **37**, 1314 (1988).

Direction of Information Flow Between Heart Rate, Blood Pressure and Breathing

Teodor Buchner, Jan J. Zebrowski, Grzegorz Gielerak and Monika Grzęda

Abstract The problem of a reliable determination of the causal precedence of physiological signals with no *a priori* knowledge is addressed. Three physiologically important signals: blood pressure variability (BPV), heart rate variability (HRV) and breathing are taken as an example. Below we introduce a novel method of analysis of physiological signals based on local linear cross-correlation. Non-stationarity of both signals is not a problem for this new method. In fact it is a requirement for the method to work. The properties of the method are confirmed by means of suitably defined test signals. The method is then applied to BPV and HRV recorded in a group of 11 patients undergoing the head-up tilt test according to a modified Westminster protocol. The results are consistent within the whole group of patients irregardless of physiological conditions and despite the fact that each of the patients exhibited a different response to tilt-test procedure. This indicates the results obtained are general. The results indicate that the breathing-related modulation of pulse pressure is one of the primary sources of variability in the cardiovascular system.

Keywords Heart rate variability · blood pressure variability · breathing · causality · correlation

1 Introduction

In the analysis of signals measured in physiological systems there often appears the important question of precedence of signals. The precedence in the time domain may be understood as a phase difference from which the direction of the information flow between studied signals may be inferred. Determination of this direction is important in many physiological cases where it is desirable to know which variability is a cause and which is an effect.

T. Buchner (✉)
Institute of Physics, Warsaw University of Technology, ul. Koszykowa 75, 00-662 Warsaw, Poland
e-mail: buchner@if.pw.edu.pl

One of the methods still widely used to determine the direction of information flow in physiological signals is frequency domain analysis and cross-spectral methods [1]. It is, however, often overlooked that the results obtained using these methods are obtained modulo 2π. This means that there is no difference between the phase equal to $-\pi/4$ and $3\pi/4$. In fact, the whole notion of information flow makes no sense if the signals are strictly periodic. If the studied systems are synchronized in their periodic regimes it cannot be said which one of them is driving and which is driven.

Among other methods used in nonlinear dynamics to determine the causality are: statistical methods such as Granger causality [2, 3] and conditional entropy [4, 5], recurrence plot methods used for the determination of the direction of coupling [6] or methods based on the Hilbert transform [7]. In the next section, a new method for the determination of causality which makes use of local nonstationarities in the signal will be introduced.

The knowledge of the direction of information flow is important when the interrelation between the heart rate variability (HRV) and the (arterial) blood pressure variability (BPV) needs to be resolved. In the BPV signal, physiologically relevant information is carried by the local minima and maxima of the blood pressure curve [8]. Thus, the quantities used here for further analysis are the diastolic blood pressure (DBP) related to the local minima and the systolic blood pressure (SBP) related with the local maxima. The difference between them is the pulse pressure (PP), which is a measure of the blood pressure added to the arterial tree due to a single heart beat.

There exist many contributions devoted either to the analysis of HRV or to the BPV separately: e.g. [9, 10]. When both variabilities are studied together, in a short time scale, their interrelation is often attributed to the action of the baroreflex loop and expressed as the "baroreflex sensitivity" (BRS). Many different methods to measure this quantity have been developed: [11–15]. On the other hand the effect of breathing on the heart rate is a well known fact [16, 17]. Thus, all the short term variabilities that are observed cannot be attributed to the result of the action of the baroreflex loop which, in addition, is a time averaged quantity. From this point of view, the name "baroreflex sensitivity" seems rather misleading [18].

The long term (circadian) behaviour of the HRV and of the BPV has also been studied [19] but there appears to be no apparent asymmetry of the information flow in such a timescale. The question of local phasic relations between both signals is outside the scope of such research.

Determination of the direction of the causal relation between the blood pressure and heart rate is important for a few reasons. Firstly it is an important modeling assumption for any reliable model of cardiovascular variability. Secondly, using the frequency domain nomenclature, where the HF range is between 0.15 and 0.4 Hz [9], it is interesting whether the primary cause of HF of BPV is the HF of HRV or not. The idea of BRS would rather suggest that, apart from the respiratory changes, the primary cause of the HRV is the BPV [15] so the question of the direction of the information flow is indeed relevant.

Convenient experimental conditions in which the above mentioned physiological signals: HRV, BPV and breathing can be measured occur during the tilt table test. The aim of this procedure is to provoke by means of physical (passive body position change) and pharmacological agents fainting in patients in order to determine the type of their response to orthostatic stress [20, 21]. The response to the activation of the various parts of the autonomous nervous system (ANS) may be examined using additional methods of provocation such as: the continuous breathing test (CBT), which activates the parasympathetic branch of the ANS [22], and the hand grip test, which activates the sympathetic branch of the ANS [23]. Because the protocol of the head-up tilt test is rich and spans many different physiological conditions, the analysis of physiological data obtained from such an examination gives a unique possibility to assess the body response to many changing physiological conditions. Concerning the information flow in the tilt test it seems interesting whether the direction or any other characteristic of the flow changes during the test or is it constant regardless of the physiological conditions.

The aim of this paper is to introduce a new method of the analysis of information flow direction: calculation of a linear cross-correlation in a sliding window. This simple method is validated on a set of test signals. Next, we determine the direction of the causal relation between the blood pressure, heart rate and breathing rate in a group of 11 patients during a head-up tilt test using a rich measurement protocol. Some new physiological results are presented.

2 Materials and Methods

We examined 11 recordings of patients who underwent the orthostatic tilt test. The recordings were selected in order to display the whole range of possible responses to the tilt test procedure. According to ACC/VASIS methodology [20, 21]: 3 of the patients exhibited cardioinhibitory response, 2 – the vasodepressive response and 6 – a mixed response.

The group consisted of 4 men and 7 women, with age 20 ± 3 and 31 ± 8 years, respectively. In 3 subjects the result of the test was positive during the passive phase of the test and 8 during the active phase of the test (description below). The patients had no evidence of cardiac arrhythmia and exhibited no ventricular or supraventricular ectopy during the tilt test recordings or before them.

2.1 Tilt-Test Protocol

Before the tilt-test procedure two stressors were applied: 10 min of the continuous breathing test and 10 min of the hand-grip test aiming to activate the parasympathetic and the sympathetic ANS, respectively. Between these tests and after them, the patient was in supine position for 20 min to ensure that all the physiological effects of stressors had died out. The tilt-test was performed according to the Westminster protocol [24] which we modified [25]. If the positive response (fainting) was not obtained within 20 min, the patient was administered sublingual nitroglycerine to provoke vasodilatation. If the test result was positive, then a beta-blocker (propranolol) was administered and the whole procedure was repeated.

2.2 Measurement Setup and Data Preconditioning

During the whole test procedure, the physiological data monitored were: a 2 channel ECG, blood pressure (Finapres) and the breathing effort signal measured by a chest belt with a stress gauge. All the variables measured were monitored by the TMS Porti device and sampled at 1,600 Hz. After completing the recording, an offline data analysis was performed including QRS detection using a simplified version of a classical algorithm [26]. The systolic and diastolic blood pressure were determined using a modified version of the algorithm described in [26]. The QRS detection did not include automatic beat classification, which was done post hoc by visual inspection of the detected QRS complexes.

Finally, all the variability signals and the breathing signal were linearly interpolated and resampled at 16 Hz to obtain an evenly sampled time series.

A typical example of the recorded signal is depicted in Fig. 1

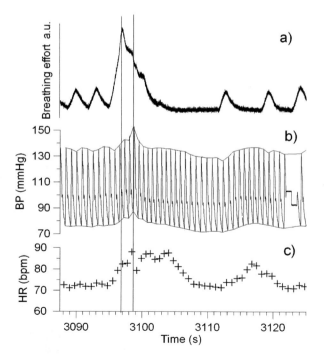

Fig. 1 Set of signals recorded in the patient 1. The *top trace* (**a**) shows the downsampled breathing effort signal. The *middle trace* depicts the measured (*thin line*) and the (downsampled) systolic and diastolic blood pressure signals. The *bottom trace* shows the HR signal. *Vertical lines* denote the nonstationarity in the breathing signal: a deep breath (the *left line*) and the response in the BP and HR curves (the *right line*)

3 Correlation of Nonstationary Signals

For periodic signals, the correlation function is periodic, thus there will be many extrema of equal size. Therefore, the time lag between the signals can only be determined modulo the period of the signal and so the direction of the information flow cannot be determined [5].

If the period of the signals considered, however, varies around a constant value, which is the case for many physiological rhythms, the envelope of the correlation function decreases monotonously as a function of the modulus of the time lag. In this case, only a few extrema in the correlation function will be obtained and the global extremum will indicate the delay. Due to the decreasing character of the correlation curve, only short time windows should be used. The timescale of the window should exceed the expected time range of the information flow process.

In many cases the direction of the information flow may be determined from the response of the correlation function to nonstationarities. In Fig. 1 it may be seen that the response to taking a deep breath is asymmetric and that the apparent direction of the information flow in this case is from breathing effort to all other signals. Such observations may provide indications which should be considered as an *a priori* knowledge.

Therefore the cause and response can be identified and the direction of information flow can be determined. Moreover, the time lag between the stimulus and the response can be measured from the position of the global maximum.

A weakness of the linear correlation function is its dependence on the actual shape of the studied signals. If the information flow direction is to be determined using this method, the response has to exhibit a waveform that is strongly correlated (in the linear sense) to the waveform of the stimulus. This conditions holds in the cases studied here, as may be seen in Fig. 1.

Another phenomenon that contributes to the information flow is that some of signals may reflect the stimulus with an opposite phase. This would be the case if the RR intervals were taken instead of the heart rate (HR). In such a case, the correlation function would be inverted (rotated around the horizontal axis) and the absolute minimum of correlation should be taken into account.

To summarize: the linear tool – cross-correlations applied to nonstationary signals may reveal information that cannot be obtained if we disregard the nonstationarity. If we expand our signal using a basis of periodic functions, we restrict ourselves to the case when the direction of information flow may only be determined modulo a period. This is a commonly known weakness of the transfer function analysis [27].

4 Local Cross-Correlation

The local cross-correlation measures the similarity between two signals: the measured signal and the reference signal. The correlations are local in such sense that only the similarity in the time scales up to the window length N is taken into account. The free parameters: the window length N and the sliding range M should be adjusted to the dominant frequency f_0 of the reference signal so that the $2\pi/f_0 < N$ and M is larger than the expected time lag between the signals.

Firstly, given a reference signal with length N, we select the initial position of the reference window in the reference signal starting from index i and a moving window of the second signal. This window will be moved from the index $i - (M - 1)/2$ to $i + (M - 1)/2$, where M is the sliding range. For each position of the moving window the linear cross-correlation is calculated [28]. This produces the correlation signal of length M. Then the reference window is shifted by N and the procedure is repeated for all positions of the reference window within the whole reference signal. A global average of all the correlation signals is then calculated yielding an average correlation signal between the reference signal and the measured signal. The midpoint of the correlation signal corresponds to the time lag between both signals equal to zero.

For each correlation signal its extrema may be determined. The positions of the extrema with respect to the midpoint of the correlation signal represent the time lags between the two signals at which they are most alike. The height of these extrema reflects the extent to which the waveforms of studied signals are alike in the sense of linear correlations, up to the selected timescale N and for a given time lag.

Only the windows in which the absolute maximum of correlation (for any lag) was greater than 0.98 should be taken into account in the average. This procedure aims to reject such windows in which the correlation between signals was too weak due to e.g. measurement errors or noise.

5 Construction of the Test Signal

In the physiological signals studied here, the nonstationarities required to measure the information flow could be observed and from visual examination of signal it could be presumed that the breathing signal is a stimulus that the other signals respond to. An example of such response to nonstationarity is shown in Fig. 1. Note that the nonstationarity in the breathing signal shown in Fig. 1 is twofold: both the amplitude and the frequency of the breathing signal changes.

In order to verify, whether the method of local correlations is able to properly describe the direction of information flow, the measurement was performed using test recordings.

In order to mimic the physiological signal, the test signal was constructed as a sine function signal with a constant bias and a random frequency: $s_i = A_0 + A_1 sin(f_n \tau \cdot i)$, where τ denotes the sampling time. The frequency was changed after one full period of the sine function. The frequency was given by: $f_n = (1+\Delta T \xi_n)/T$ where T is the mean period of the sine function, ΔT represents the magnitude of the period changes and ξ_n are uniformly distributed random variables in the range $[-1;1]$ and n denotes the index of the period.

Direction of Information Flow

This type of nonstationarity will be termed *nonstationarity in frequency*. The changing frequency of the signal (which is almost periodic) is observed in the heart rate as well as in the breathing rate.

Moreover at random positions in the signal we introduced a *nonstationarity in amplitude*. The amplitude was changed in such a way that for one whole period of the sine function the amplitude was increased stepwise: it was multiplied by the factor $K > 1$. The moments at which this nonstationarity occurred were chosen randomly and the distance between them Δn satisfied $\Delta n = n_0 + n_1 \cdot \xi_n$. Such a nonstationarity mimics the type of nonstationarity observed in the breathing signal.

Note that if the subsequent periods do not change much, the neighbouring maxima may have equal or nearly equal height which may be misleading. On the other hand, it should be possible to select a number of windows in signal where the periods do change much and determine the unique direction and the time lag of the information flow from these windows. Thus, the procedure is more robust if the nonstationarity in frequency is stronger.

The parameters used in the test signals studied in this work were:

- the mean period of the sine function $T = 160$ and the magnitude of the period changes $\Delta T = 0.2T$.
- the mean number of periods between nonstationarities $n_0 = 10$ and the magnitude of changes $n_1 = 4$.
- the amplitude of stationary periods $A_1 = 20$
- the nonstationary amplitude increase factor $K = 1.2$
- the mean value of the test signal $A_0 = 100$

The test signal in such settings had an amplitude range 80–120 – similar to that of the blood pressure for the sake of convenience, but the method is not sensitive to the values of A_0 and A_1. All the subsequent time intervals in following text will be expressed as multiples of T.

Two copies of the same test signals were prepared with a suitable time shift between them. The time shift could be changed in order to determine if the method can reliably detect the value and sign of the time lag.

6 Results

6.1 Local Cross-Correlations of the Test Signals

For the purpose of the test procedure three signals: A, B, C and D were generated. The time shifts of two test signals B and C with respect to the reference signal A were $T_{AB} = -T/8$ for signal B and $T_{AC} = -1.25\,T$ for signal C. The signal D had $T_{AD} = -1.25\,T$ but contained no amplitude nonstationarities. All the test signals are shown in Fig. 2.

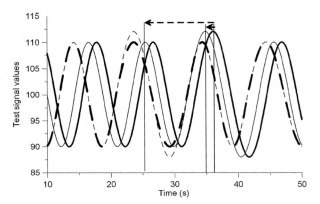

Fig. 2 Test signals A – *thick solid line*, B – *thin solid line*, C – *thin dashed line*, D – *thick long-dash line*. The time lag T_{AB} is marked with the *solid line arrow*. The time lag T_{AC} is marked with the *dashed line arrow*. The *vertical lines* denote the maxima of the nonstationarities

For the test signals A, B and C, local cross-correlations were calculated with the window length $N = T$ and the sliding range $M = 2T$. The correlation curve averaged over all positions of the reference window is shown in Fig. 3 as a function of the time lag.

The shape of the local cross-correlation curve is an oscillatory function with the amplitude decreasing as a function of the time lag. The period of these oscillations is equal to T and the amplitude decrease is caused by the random variations of the sine frequency that cause an incoherence of signals. The length of time lag $T_{AC} > |T|$. Despite the fact that the amplitude of correlation generally decreases, the global maximum is clearly visible (thick dashed curve in Fig. 3). Note that this maximum is larger than the one with the smaller value of $|T|$ for the same curve. This global maximum is positioned at $-T_{AC}$. We see then that, due to the presence of nonstationarities, the time lag of the absolute maximum indeed reflects the time lag between the studied signals. Note also that the shape of the correlation curve is identical for the signals C and D. This means that, for the test signal studied, the shape of the correlation curve is entirely determined by the nonstationarity in frequency. The irregularity of the signal is an important feature that enables us to reliably determine

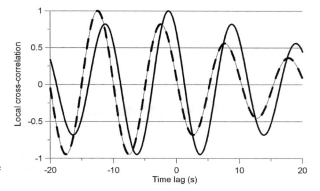

Fig. 3 Local cross-correlations for test signal B (*thick solid line*), C (*thick dashed line*) and D (*thin solid line*) with reference signal A. The window length $N = T$, the sliding range $M = 2T$ and the threshold of correlation was 0.98. Note that two of the signals overlap

the time lag. Note also that the window length N was set to the mean period T of the quasiperiodic signal. If the value of N is set to $N > T$ the rate at which the cross-correlation function vanishes will be even faster. On the other hand setting $N = T$ is enough to obtain the incoherence and the decay of correlation.

6.2 Local Cross-Correlations of the Physiological Signal

Having validated the procedure, we calculated local cross-correlations for the physiological signal.

When the breathing signal is used as reference, the value of the sliding range M is set to the number of samples equivalent to 6.5 s that is two periods of the breathing signal (taking the mean physiological breathing rate: 18 beats per min as reference).

A typical result is shown in Fig. 4. It may be seen that the correlations of all the measured signals have a positive time lag towards breathing. This means that the information flow is from breathing to these signals.

Figure 5 depicts the group averaged correlation curves. It may be seen that the relative height of certain maxima has changed but the positions and the signs of time lags remain the same in spite of the wide range of physiological conditions in the averaged group. The standard deviation in the group was also analyzed. It was found not to exceed 0.21 for any time lag within the sliding range and for any measured signal.

Figure 6 shows the global extrema of local cross-correlations that were obtained for each individual recordings and for the group average. There are two cases (pts 8, 11) where the absolute value of the extremum of correlation is below 0.5 for some measured signals, which suggests that the correlation is weak.

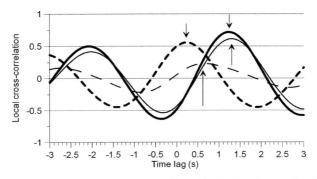

Fig. 4 Correlation signal for the patient 1. The breathing effort is the reference signal and the traces show correlations for the different measured signals: heart rate (*thick dashed line*), systolic (*thin solid line*), diastolic (*thin dashed line*) and pulse pressure (*thick solid line*). Starting from the *left* of the figure, the *arrows* denote the global extrema of the correlation functions: for HR, diastolic BP, systolic BP and pulse BP

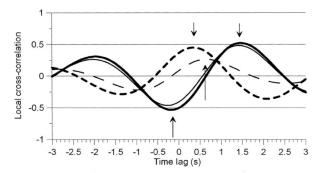

Fig. 5 Group averaged correlation signal. The breathing effort is the reference signal and the traces show correlations for the different measured signals: heart rate (*thick dashed line*), systolic (*thin solid line*), diastolic (*thin dashed line*) and pulse pressure (*thick solid line*). Starting from the left of the figure, the arrows denote the global extrema of the correlation functions: for pulse BP, HR, diastolic BP and systolic BP

Next, we analyzed the correlations with the heart rate signal taken as reference. The sliding range M was set to the number of data samples equivalent to 4 s. and the value of N was set to the number of samples equivalent to 2 s. A typical result is shown in Fig. 7.

From the analysis of the correlation curve it may be seen that there is a feed-forward type relation from HR to the systolic pressure and pulse pressure at the following RR interval.

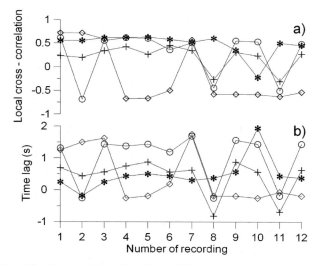

Fig. 6 The values (**a**) and the positions (**b**) of the global extrema of the local cross-correlations for all the patients (1–11) and the group average (recording 12). The reference signal is the breathing effort and the traces show cross-correlations for: the heart rate (∗), the systolic (o), the diastolic (+) and the pulse (◊) blood pressures

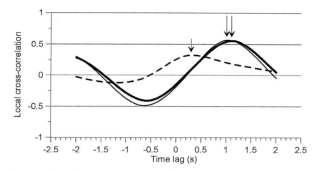

Fig. 7 Correlation signal obtained from patient 1. The reference signal is heart rate and the traces show correlations with different measured signals: systolic pressure (*thick solid line*), diastolic pressure (*dashed curve*) and pulse pressure (*thin solid line*). Starting from the *left* of the figure, the *arrows* denote the global extrema of the correlation functions: for diastolic BP, pulse BP and systolic BP

7 Discussion

The results described above seem to show a rather clear picture of the physiological information flow. The strongest relation with breathing is that of the pulse pressure, which is well visible for almost all patients in Fig. 6a when the absolute values of the extrema are plotted (not shown). Despite the sign of the correlation (i.e. is it a minimum or a maximum), the extremum of correlation of the pulse pressure has the largest absolute value. Therefore, we may conclude that it is the pulse pressure which is causally the strongest related to the breathing effort. Of only slightly smaller magnitude is the correlation between the systolic blood pressure, the heart rate and the breathing effort.

The strong relation between the blood pressure and the breathing effort may be understood in view of the breathing-related changes in haemodynamics described e.g. by Toska et al. [29]. They showed that the blood flow in the large vessels (especially the veins) located within the chest is strongly affected by the intrathoracic pressure changes induced by the breathing effort. This introduces a modulation in the blood flow which is transferred to the pulse pressure [29]. This pulse pressure modulation is transferred into the systolic blood pressure modulation.

The systolic blood pressure has its global extremum at 1.5 s in 8 of 11 cases. This reveals the strong feedforward relation between the breathing effort and the systolic pressure.

Concerning the time-lag between the breathing effort and pulse pressure the situation is not quite clear. In 4 of 11 patients (1, 2, 3, 7) the highest absolute value of the cross-correlation for pulse pressure coincides with the extrema of the systolic pressure. The pulse pressure was determined at the time of the subsequent systolic peak, therefore the time lag is identical for both the systolic pressure and pulse pressure. This reveals the feedforward relation between the breathing effort and the pulse pressure, which is practically transferred to the systolic pressure in a 1:1 manner.

Therefore, in these cases, we see a clear information flow from the breathing effort to the pulse pressure and systolic pressure.

In other patients, the minimum at -0.5 s occurred to be the global extremum. This does not necessarily mean that the information flow is reversed. It may just as well indicate that the breathing related information has some other source e.g. venous return, which affects the pulse pressure. If the phase of such information is reversed it would give a modified image of the information flow. The venous return is known to have a strong influence on pulse pressure [30] so the above finding is plausible. The pulse pressure depends also on HR (c.f. Fig. 7) which also depends on breathing. This type of indirect relation may also affect the correlation.

Note also that, as may be seen in Fig. 5 in pulse pressure, the absolute values of the maximum ($\tau = 1.6$ s) and the minimum ($\tau = -0.2$ s) of the local cross-correlation function differ only by 1.6% therefore the effect of the additional source of information is not strong.

Another observation ensures that the information flow is directed from the breathing effort to pulse pressure and systolic pressure and not vice versa. It is the kind of response to the amplitude type nonstationarity i.e. a spontaneous deep sigh. Observation of such clear cut responses (like the one shown in Fig. 1) shows convincingly that the basic mechanism which determines the direction of information flow is the breathing modulation of pulse pressure. All the results that are different from that show effects of various physiological origin.

From the above example it may be seen that, although the time lag determined by the method of the local cross-correlations is not given modulo one period, still some caution should be taken during the interpretation of this quantity. The problem arises in case when many local extrema, situated at distant time lags have values of local cross-correlation that only slightly differ from each other. This problem does not change the value of correlation significantly but may introduce a large error into the measurement of the time lag. Therefore it is valuable either to have additional evidence concerning the information flow or limit the set of local windows. For instance one could include in the statistics only such windows of measured signal where the difference between the value of the first and the second extremum exceeds some threshold.

Analysing the correlation for the heart rate, one may see that HR is nearly in phase with breathing. Within 300 ms after inspiration the HR is the largest. This is consistent with other observations of the respiratory sinus arrhythmia [31]. The minimum value of HR is around 1.875 s after the inspiration (related with the time lag 0). Note also that the value of HR is associated with the point in time at which the RR interval ends. Therefore the interval over which the RR is measured begins earlier than 1.875 s. This means that the HR and the systolic/pulse blood pressure are nearly in counterphase. The respiratory modulation of the heart rate is large as the group of patients is young and exhibits a strong vagal tone.

Concerning the correlation for the diastolic blood pressure, note that the relative height of this curve is the smallest (<0.4) and that the highest local maximum is located at 0.625 s showing also a feedforward relation type from breathing to the diastolic blood pressure.

We may conclude that *the modulation by breathing of the heart rate mediated by the parasympathetic branch of the ANS seems to have a stabilizing role*. This observation is supported by the fact that the correlations between the systolic/pulse blood pressure, heart rate and breathing are the strongest whereas the correlations between the diastolic blood pressure and breathing are the weakest among all studied. This also sheds the new light on the open problem of the origins of the respiratory sinus arrhythmia [32]. According to [33], in an upright position the respiratory sinus arrhythmia is able to stabilize the breathing-related fluctuations in the blood pressure. It is interesting that this result has been here confirmed in a noninvasive way using the time series analysis.

The correlations obtained with heart rate as reference signal show another interesting effect: the systolic pressure of the following heart beat is related to the preceding heart rate. This may be understood in the sense that during the shorter diastolic interval the blood pressure was able to decay to a higher value than during the slower heart rate.

8 Conclusions

1. The direction of the information flow between two signals may be deduced using a local linear correlation function. The condition for this to be feasible is that the signal must be nonstationary in either the frequency or in the amplitude.
2. Our findings support the observation made by Toska et al. that the stroke volume and, hence, the systolic pressure as well as the pulse pressure are strongly modulated by the breathing rhythm and that this modulation is of a predominantly haemodynamical (and not neurological) nature [29]. We show that this effect may be observed through a medically noninvasive analysis of the local correlations between blood pressure and the breathing rhythm. It seems that this modulation is one of the important sources of the variability of the systolic blood pressure in healthy humans, in physiological conditions. Our findings also suggest that there is an alternative route through which breathing affects the system which is probably related to the dynamics of the vascular bed.
3. There seems to be a simple relation between the shortening of the RR interval and the increase of the diastolic blood pressure.

References

1. Wichterle D, Melenovsky V, Simek J, et al. (2000) Cross-spectral analysis of heart rate and blood pressure modulations. PACE 23(9):1425–1430.
2. Nollo G, Faes L, Porta A, et al. (2005) Exploring directionality in spontaneous heart period and systolic pressure variability interactions in humans: implications in the evaluation of baroreflex gain. Am J Physiol Heart Circ Physiol 288:H1777–H1785.
3. Marinazzo D, Pellicoro M, and Stramaglia S. (2006) Nonlinear parametric model for Granger causality of time series. Phys Rev E 73:066216.

4. Hoyer D, Frank B, Goetze C, et al. (2007) Interactions between short-term and long-term cardiovascular control mechanisms – a new aspect in risk stratification. Chaos 17:015110-1-015110-8.
5. Schreiber T. (2000) Measuring information transfer. Phys. Rev. Lett. 85:461.
6. Romano Blasco MC. (2004) Synchronization analysis by means of recurrences in phase space. Dissertation, University of Potsdam. http://opus.kobv.de/ubp/volltexte/2005/184/pdf/ROMANO.PDF, cited 3 Oct 2007.
7. Rosenblum MG, Cimponeriu L, Bezerianos A, et al. (2002) Identification of coupling direction: application to cardiorespiratory interaction. Phys Rev E. 65(4):041909.
8. Guyton AC, Hall JE. (2006) Textbook for Medical Physiology, ed. 11, Elsevier, Philadelphia.
9. Task Force of the European Society of Cardiology and The North American Society of Pacing and Electrophysiology. (1996) Guidelines—Heart rate variability: standards of measurement, physiological interpretation, and clinical use. Eur Heart J 17:354–381.
10. Fratolla A, Parati G, Cuspidi C, et al. (1993) Prognostic value of 24-hour blood pressure variability. J Hypertens 11:1133–1137.
11. Laude D, Elghozi J-L, Girard A, et al. (2004) Comparison of various techniques used to estimate spontaneous baroreflex sensitivity (the EuroBaVar study). Am J Physiol Regul Integr Comp Physiol 286:R226–R231.
12. Parati G, Di Rienzo M, Bertinieri G, et al. (1988) Evaluation of the baroreceptor-heart rate reflex by 24-hour intra-arterial blood pressure monitoring in humans. Hypertension 12:214–222.
13. Pagani M, Lombardi F, Guzzetti S, et al. (1986) Power spectral analysis of heart rate and blood pressure variabilities as a marker of sympatho-vagal interaction in man and conscious dog. Circ Res 59:178–193.
14. Robbe HWJ, Mulder LJM, Ruddel H, et al. (1987) Assessment of baroreceptor reflex sensitivity by means of spectral analysis. Hypertension 10:538–543.
15. Westerhof BE, Gisolf J, Stok WJ, et al. (2004) Time-domain cross-correlation baroreflex sensitivity: performance on the EUROBAVAR data set. J Hypertens 22:1–10.
16. Grossman P, van Beek J, Wientjes C (1990) A comparison of three quantification methods for estimation of respiratory sinus arrhythmia. Psychophysiology 27(6):702–714.
17. Hayano J, Yasuma F, Okada A, et al. (1996) Respiratory sinus arrhythmia: a phenomenon improving pulmonary gas exchange and circulatory efficiency. Circulation 94(4):842–847.
18. Wesseling KH. (2002) Baroreflex sensitivity, an elusive number. http://www.finapres.com/files/articledownloads/full-wesseling-2002.pdf, cited 3 Oct 2007.
19. Mancia G, Ferrari A, Gregorini L, et al. (1983) Blood pressure and heart rate variabilities in normotensive and hypertensive human beings. Circ Res 53(1):96–104.
20. Benditt D, Ferguson D, Grubb B, et al. (1996) Tilt table testing for assessing syncope: ACC expert consensus document. J Am Coll Cardiol 28:263–275.
21. Sutton R, Petersen M, Bringole M, et al. (1992) Proposed classification for tilt induced vasovagal syncope. Eur J Card Pacing Electrophysiol 2:180–183.
22. Bannister R, Mathias CJ (eds). (1992) Autonomic Failure: A Textbook of Clinical Disorders of the Autonomic Nervous System, Oxford University Press, Oxford.
23. Khurana RK, Setty A. (1996) The value of the isometric hand-grip test-studies in various autonomic disorders. Clin Auton Res 6:211–218.
24. Kenny RA, Ingram A, Bayliss J, Sutton R. (1986) Head-up tilt: a useful test for investigating unexplained syncope. Lancet 8494:1352–1355.
25. Gielerak G, Makowski K, Cholewa M. (2005) Prognostic value of head-up tilt test with intravenous beta-blocker administration in assessing the efficacy of therapy in patients with vasovagal syncope. Ann Noninv Electrocard 10(1):65–72.
26. Pan J, Tompkins WJ. (1985) A real-time QRS detection algorithm. IEEE Trans Biomed Eng, BME-32(3):230–236.
27. Saul JP, Berger RD, Albrecht P, et al. (1991) Transfer function analysis of the circulation: unique insights into cardiovascular region. Am J Physiol 261:H1231–H1245.

28. Oppenheim AV, Schafer RW. (1975) Digital Signal Processing, Prentice-Hall, Englewood Cliffs, 1975.
29. Toska K, Eriksen M. (1993) Respiration-synchronous fluctuations in stroke volume, heart rate and arterial pressure in humans. J Physiol (London) 472:501–512.
30. Guyton AC. (1995) Determination of cardiac output by equating venous return curves with cardiac response curves. Physiol. Rev. 35:123–129.
31. Hirsch JA, Bishop B. (1981) Respiratory sinus arrhythmia in humans: how breathing pattern modulates heart rate. Am J Physiol 241 (Heart Circ Physiol 10):H620–H629.
32. Yasuma F, Hayano J-i. (2004) Respiratory sinus arrhythmia: when does the heart beat synchronize with respiratory rhythm? Chest 125:683–690.
33. Elstad M, Toska K, Chon Ki H, et al. (2001) Respiratory sinus arrhythmia: opposite effects on systolic and mean arterial pressure in supine humans. J Physiol 536(1):251–259.

Part II
Cardiovascular Physics: Modelling

The Mathematical Modelling of Inhomogeneities in Ventricular Tissue

T.K. Shajahan, Sitabhra Sinha and Rahul Pandit

Abstract Cardiac arrhythmias such as ventricular tachycardia (VT) or ventricular fibrillation (VF) are the leading cause of death in the industrialised world. There is a growing consensus that these arrhythmias arise because of the formation of spiral waves of electrical activation in cardiac tissue; unbroken spiral waves are associated with VT and broken ones with VF. Several experimental studies have been carried out to determine the effects of inhomogeneities in cardiac tissue on such arrhythmias. We give a brief overview of such experiments, and then an introduction to partial-differential-equation models for ventricular tissue. We show how different types of inhomogeneities can be included in such models, and then discuss various numerical studies, including our own, of the effects of these inhomogeneities on spiral-wave dynamics. The most remarkable qualitative conclusion of our studies is that the spiral-wave dynamics in such systems depends very sensitively on the positions of these inhomogeneities.

Keywords Excitable media · ventricular fibrillation · spiral turbulence · cardiac arrhythmias · spatiotemporal chaos

1 Introduction

Mammalian hearts are electromechanical pumps in which the propagation of electrical activation waves through cardiac tissue triggers the mechanical contraction of the walls of the atria or ventricles. The walls of these chambers are made of tissue, which is an excitable medium that supports the passage of regular contraction waves across it. This excitability depends crucially on the states of cardiac cells that make up the tissue. In its quiescent state, such a cell maintains a potential difference of $\simeq 85$ mV across the cell membrane. If the cell is stimulated beyond a certain threshold potential ($\simeq 60$ mV), voltage-gated ion channels on the membrane allow Na$^+$

T.K. Shajahan (✉)
Department of Physics, Centre for Condensed Matter Theory, Indian Institute of Science, Bangalore 560012, India,
e-mail: shajahan@physics.iisc.ernet.in, shajahan.tk@gmail.com

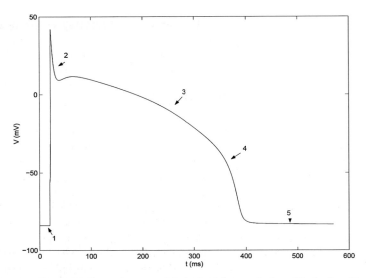

Fig. 1 An example of a cardiac action potential obtained from a single cell in the Luo-Rudy model showing (1) the action potential upstroke or rapid depolarization, (2) the rapid repolarization, (3) the plateau stage, (4) the final stage of repolarization, and (5) the resting state, in a plot of transmembrane potential V versus time t

ions to enter the cell. Thus the cell is rapidly depolarized; it repolarizes partially for $\simeq 10$ ms, and then enters a plateau state in which ion channels, for Ca^{++} and K^+ ions, are activated. This leads to the repolarization of the cell and it returns eventually to its quiescent state. The repolarization phase lasts much longer than the initial depolarization phase (hundreds of ms compared to a few ms). The depolarization-repolarization response is referred to as the action potential. The action potential duration (APD) in a human ventricular cell is typically $\simeq 200$ ms; a typical plot of an action potential is shown in Fig. 1. During the period of repolarization and slightly after it, a cardiac cell cannot be easily excited again and is said to be in a refractory state. The coupling between cardiac cells results in the propagation of excitation from one cell to the other.

In a normal heart, the regular rhythmic contraction of the atria and ventricles pumps blood through the body. Cardiac arrhythmias disturb this regular rhythm; the most dangerous of these are ventricular tachycardia (VT) and ventricular fibrillation (VF). During an episode of VT the rate at which a patient's heart beats increases to more than 100 beats per min; an electrocardiogram (ECG) shows a roughly periodic trace but one that is considerably faster than the normal sinus rhythm; if VF occurs then, not only does the rate of heart beats increase, but the ECG trace seems quite chaotic. (An introduction to ECGs that is accessible to nonspecialists can be found at Ref. [1]). There is growing consensus that VT is associated with the formation of a single spiral wave of electrical activation in ventricular tissue whereas VF is believed to arise when such spiral waves break leading to spiral turbulence. Voltage-sensitive dyes and CCD cameras can be used to record the propagation of the such spiral waves of excitation in cardiac tissue [2–5]. Indeed spiral waves are ubiquitous in excitable media, in which subthreshold excitations decay rapidly but suprathreshold excitations lead to a response like the action potential described above. Cardiac tissue is one example of an excitable medium. Such excitable media are modeled by coupled, nonlinear, partial differential equations that describe the spatiotemporal evolution of concentration fields; in suitable parameter ranges these models can show regular spiral waves or spiral turbulence that is accompanied by spatiotemporal chaos [6]. In the next Section we will discuss two such models for cardiac tissue.

There are several mechanisms for the onset of spiral waves in excitable media. One mechanism relies on electrophysiological inhomogeneities, set up in the tissue because of the passage of the spiral wave; i.e., the refractory tissue in the wake of an activation wave can act as a temporary obstacle in the medium. Such reentry is referred to as functional reentry. Another mechanism invokes the presence of permanent inhomogeneities in cardiac tissue, e.g., inexcitable cells, such as scar tissue caused by a previous cardiac arrest or ischaemia. Excitation waves can move around such permanent inhomogeneities; the resulting reentry is referred to as anatomical reentry and it can lead to a spiral wave that moves around the obstacle.

Tissue heterogeneities, which can occur at several length scales (from microns to mm), can be of various types. A small percentage of cardiac tissue is fibrotic and non-excitable; this percentage can increase significantly (up to 30–40%) during aging, after a myocardial infarction, or in cardiac myopathies. A myocardial infarction can leave slightly bigger obstacles of poorly excitable regions (of the order of mm). Major arteries in the heart can also act as obstacles in the path of a propagating wave. We concentrate on inhomogeneities that are large enough that partial-differential-equation models for cardiac tissue are valid.

Clearly the effects of inhomogeneities, functional and anatomical, on wave propagation and spiral-wave formation and break up in cardiac tissue is an important question that has a direct bearing on life-threatening cardiac arrhythmias. Not surprisingly then various experimental and theoretical studies have been undertaken to address this question. We begin with a very brief overview of studies that have concentrated on understanding the dynamics of spiral waves in two-dimensional cardiac tissue in which some inhomogeneities are introduced systematically.

Experiments conducted on ventricular tissue with a single obstacle have shown that small obstacles do not affect spiral waves but, as the obstacle size increases, a spiral wave can get attached to the obstacle and then start moving around it [7]. Davidenko et al. [8] have found that, in one case, an artificially induced spiral wave moved away from its site of origin in their cardiac-tissue preparation because of an obstacle. By contrast, other studies [9–12] have shown that an obstacle, in the path of a moving spiral wave, can break it and lead to many competing spiral waves. Recent experiments by Hwang et al. [13] have suggested that multistability of spirals with different periods in the same cardiac-tissue preparation can arise because of the interaction of spiral tips with small-scale inhomogeneities.

It has also been found in numerical simulations of two-dimensional models for cardiac tissue that such obstacles can convert a meandering (detached) spiral wave into a stable (attached) one [14–16]. A naturally occurring obstacle like papillary muscle can also play the role of an attaching structure during VT [17].

A systematic numerical study has been carried out for a circular obstacle (by creating a hole in the simulation domain for the Luo-Rudy model). This study shows that, if the hole is large enough, a spiral wave can get anchored to it; but, as the radius of the hole is decreased, there is a transition from periodic to quasiperiodic motion of the spiral wave via a Hopf bifurcation, and eventually a transition to spiral-wave breakup and spatiotemporal chaos [14].

Starobin et al. [18] have studied, numerically and analytically, the interaction of spiral waves with piece-wise linear obstacles. They find that if the excitability of the medium is high, the wave moves around the obstacle boundary, rejoins itself and then proceeds as if it had not encountered any obstacles in its path. However, if the excitability is low, the two ends of this wavefront are unable to join, so two free ends survive, curl up and develop into two spiral waves, which in turn can break up again. This study also finds that, apart from the excitability of the medium and the local curvature of the wave front, the shape of the obstacle also affects the attachment of the spiral to the obstacle.

Anchoring helps in the control of such spiral waves. In the anchored condition the spiral wave is considerably less dangerous than a broken spiral wave for the former is associated with VT whereas the latter is associated with VF. Furthermore, another numerical study has shown that [19] a single anchored spiral wave can be easily unpinned and removed from the simulation domain by applying a weak, uniform electric field (less than 0.5 V/cm). Such a method is very useful because one does not need to know the exact location of the spiral tip to eliminate it from the domain.

A numerical study of a FitzHugh-Nagumo-type model has shown that spiral break up can be suppressed by introducing a large fraction of non-excitable cells, distributed randomly in the simulation domain [20,21]. In the following sections we give an overview of our numerical studies of spiral-wave dynamics in the presence of obstacles in two models of ventricular tissue. Our goal has been to design studies to bring out the dependence of such dynamics on the position, size, and shapes of such obstacles. Some details of this work have appeared in Ref. [22], where we have shown that spiral-wave dynamics depend sensitively on all these parameters; we also present some new results here for the interactions of spiral waves with two inhomogeneities. Section 2 is devoted to a description of the models and our numerical schemes. Section 3 contains an overview of our results. Section 4 concludes this paper with a discussion of the implications of our studies for cardiac arrhythmias and their control.

2 Models

In this section we give a brief description of the two mathematical models for ventricular tissue that we use in our studies. These are the Panfilov model [23, 24] and the Luo-Rudy I (LRI) model [25]. The former belongs to simplified models of the

FitzHugh-Naguma type; the latter is a detailed physiological models based on the Hodgkin-Huxley formalism. Such detailed models account for the ionic currents that are transported through voltage-gated ion channels in cardiac tissue, so their numerical simulation is computationally expensive. Simulations of simplified models like that of Panfilov [23,24], are not as demanding from the point of view of computational resources; however, they often suffice to obtain results that are qualitatively similar to those that follow from studies of detailed models like LRI [14, 26–29]. As we will see below, the ways in which spiral waves appear, break up, and interact with obstacles, and the way in which spiral turbulence can be controlled in the Panfilov model are qualitatively similar to those in the LRI model. Our discussion is based predominantly on recent studies [22, 26–29] that we have carried out in our group; our earlier studies of the control of spiral turbulence in these models can be found in Refs. [6, 29].

The Panfilov model consists of a partial differential equation (PDE) for the transmembrane potential $V(x,t)$ coupled to an ordinary differential equation for the recovery variable $g(x,t)$ at the point x and time t. All information about the ion channels is lumped into g, which is related to the membrane conductance. The Panfilov-model equations are

$$\partial V/\partial t = \nabla^2 V - f(V) - g,$$
$$\partial g/\partial t = \varepsilon(V, g)(kV - g). \qquad (1)$$

The initiation of action potential arises because of the following piecewise-linear form for $f(V)$: $f(V) = C_1 V$, for $V < e_1$, $f(V) = -C_2 V + a$, for $e_1 \leq V \leq e_2$, and $f(V) = C_3(V - 1)$, for $V > e_2$. We use physically appropriate parameters given in Refs. [23, 24]: $e_1 = 0.0026$, $e_2 = 0.837$, $C_1 = 20$, $C_2 = 3$, $C_3 = 15$, $a = 0.06$, and $k = 3$. The function $\varepsilon(V, g)$ determines the dynamics of the recovery variable: $\varepsilon(V, g) = \varepsilon_1$ for $V < e_2$, $\varepsilon(V, g) = \varepsilon_2$ for $V > e_2$, and $\varepsilon(V, g) = \varepsilon_3$ for $V < e_1$ and $g < g_1$, with $g_1 = 1.8$, $\varepsilon_1 = 1/75$, $\varepsilon_2 = 1.0$ V, and $\varepsilon_3 = 0.3$.

We will also show representative results from our simulations of the Luo-Rudy I (LRI) model, which is based on the Hodgkin-Huxley formalism. It accounts for 6 ionic currents [25] (e.g., Na^+, K^+, and Ca^{2+}) and 9 gate variables for the voltage-gated ion channels that control ion movement across the cardiac cell membrane. In the resting state there is a potential difference of $\simeq 84$ mV across this cell membrane. If a stimulus raises this above -60 mV, Na^+ channels open, resulting in an action potential (Fig. 1). Cells in the LRI model are coupled diffusively; thus one must solve a PDE (see Appendix) for the transmembrane potential V; the time evolution and V dependence of the currents that appear in this PDE are given by 7 coupled ordinary differential equations (ODEs) [22, 25].

We restrict ourselves to two-dimensional studies here and refer the reader to Ref. [22] for our three-dimensional simulations. For both the Panfilov-model and LRI PDEs we use a forward-Euler method in time t and a finite-difference method in space that uses a five-point stencil for the Laplacian. Our spatial grid consists of a square lattice with side L mm. For the Panfilov model we use temporal and spatial steps $\delta t = 0.022$ and $\delta x = 0.5$, respectively, and $L = 200$ mm. We take the dimensioned time [23, 24] T to be 5 ms times dimensionless time and 1 spatial unit to be 1 mm; the dimensioned value of the conductivity constant D is $2\,\text{cm}^2/\text{s}$. For the LRI model we use $\delta t = 0.01$ ms, $\delta x = 0.0225$ cm, and $L = 90$ mm.

We initialize Eq. (1) at $t = 0$ with the following broken wave-front (Fig. 2A and 2B): $g = 2$ for $0 \leq x \leq L$ and $0 \leq y \leq \frac{L}{2}$, and $g = 0$ elsewhere (L is the linear system size); $V = 0$ everywhere except for $y = \left(\frac{L}{2} + 1\right)$ and $0 \leq x \leq \frac{L}{2}$, where $V = 0.9$. We observe spiral turbulence if $L > 128$ mm and $\varepsilon_1 = 1/75$; as ε_1 decreases, the spiral pitch decreases and, eventually, broken spirals are obtained; e.g., if $\varepsilon_1 \simeq 0.01$, a state containing broken spirals is observed (Fig. 2B and 2C). However, if $\varepsilon_1 \geq 0.02$, Eq. (1) displays rigidly rotating spiral as discussed in Pandit et al. [6]. Furthermore the breakup of spirals, for $\varepsilon_1 > 1/75$, is associated with the onset of spatiotemporal chaos [6]; spatiotemporally chaotic behaviour is, strictly speaking, a transient, but the duration τ_L of this transient increases with the size L of the system; for $t > \tau_L$, a quiescent state with $V = g = 0$ is obtained. If $\varepsilon_1 > 1/75$ and $L > 128$, τ_L is sufficiently large and we find a nonequilibrium statistical steady with spatiotemporal chaos [6, 29], i.e., the number of positive Lyapunov exponents increases with L, as does the Kaplan-Yorke dimension D_{KY}. Thus VF in the Panfilov model is associated with spiral turbulence and spatiotemporal chaos, initiated by the breakup of spiral waves. Similar spiral turbulence has been observed in more

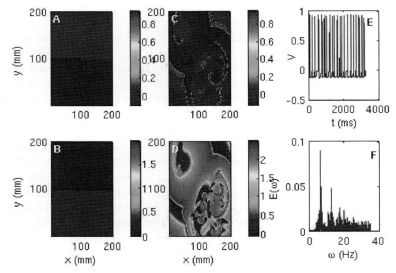

Fig. 2 Spiral-wave break up in the Panfilov Model: Pseudocolour plots of the initial conditions used in our simulation for the fields (**A**) V and (**B**) g. This initial condition leads to the formation of a spiral wave that breaks down eventually leading to spiral turbulence for (**C**) V and (**D**) g fields after $t = 2{,}750$ ms. (**E**) The time series of V from the point (100 mm, 100 mm) and (**F**) the associated power spectrum [computed after eliminating transients, i.e., the first 50,000 time steps (5,500 ms) from a time series of 311 424 time steps (34256.64 ms)]; this broad-band power spectrum is characteristic of a chaotic state

The Mathematical Modelling of Inhomogeneities in Ventricular Tissue 57

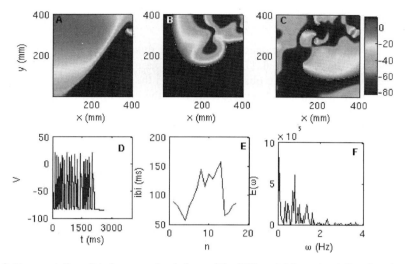

Fig. 3 Representative plots from our simulations of the LRI model for a simulation domain with $L = 90$ mm. Pseudocolour plots of the transmembrane voltage V: (**A**) initial condition ($t = 0$); (**B**) after 600 ms; and (**C**) after 1,000 ms. Note that spiral wave has broken in (**B**) and (**C**). (**D**) The local time series of V, taken from the point (45 mm, 45 mm); (**E**) the inter-beat interval (IBI), between two successive spikes in (**D**), versus the spike number n; and (**F**) the power spectrum associated with the time series of (**D**). The fluctuations of the IBI and the broad-band power spectrum are signatures of the underlying spiral turbulence

realistic models for ventricular tissue [14, 22] like the Luo-Rudy I model; to the best of our knowledge, Lyapunov spectra and D_{KY} have not been obtained for these models.

The initial condition for transmembrane potential used in LRI models is shown in Fig. 3A; in the absence of any obstacle in the medium this develops as shown in Fig. 3B and 3C.

We model conduction inhomogeneities by setting $D = 0$ in regions with obstacles; in all other parts of the simulation domain D is a constant (1 for the Panfilov model and between 0.0005 cm^2/s and 0.001 cm^2/s for the LRI model). In the studies we present here the obstacle is taken to be a square of side l mm, with 10 mm $\leq l \leq 40$ mm. For both models we use no-flux (Neumann) boundary conditions on the edges of our simulation domain and on the boundaries of the obstacle (or obstacles).

3 Obstacles in the Medium

We have carried out several studies designed to elucidate the effects of conduction inhomogeneities in the Panfilov and LRI models. These are described in detail in Ref. [22]. Here we give an overview of these results and present some new studies that we have carried out on the interplay of two conduction inhomgeneities with spiral waves.

We first examine the dependence of spiral-wave dynamics by fixing the position of the obstacle (cf., Ikeda et al. [7] for similar experiments) and changing the obstacle size: We place a square obstacle of side l in the Panfilov model in a square simulation domain with side $L = 200$ mm. With the bottom-left corner of the obstacle at (50 mm, 100 mm) spiral turbulence (ST) persists if $l \leq (40 - \Delta)$ mm; ST gives way to a quiescent state (Q) with no spirals if $l = 40$ mm, and eventually to a state with a single rotating spiral (RS) anchored at the obstacle if $l \geq (40 + \Delta)$ mm. We have taken l from 2 mm to 80 mm in steps of $\Delta = 1$ mm. Thus we see a clear transition from ST to RS, with these two states separated by a state Q with no spirals.

The final state of the system depends on its size and also on where the obstacle is placed with respect to the tip of the initial wavefront. Even a small obstacle, placed near this tip [obstacle with $l = 10$ mm at (100 mm, 100 mm)], can prevent the spiral from breaking up; but a bigger obstacle, placed far away from the tip [obstacle with $l = 75$ mm at (125 mm, 50 mm)], does not affect the spiral.

In Ref. [22] we have explored in detail how the position of the obstacle changes the final state; here we give an overview of these results only for the two-dimensional case. One of the remarkable conclusions of our study is that the final state of the system depends sensitively on the position of the obstacle. This is illustrated clearly in Figs. 4, 5, 6 for the Panfilov model and Figs. 7, 8, 9 for the LRI model. As these

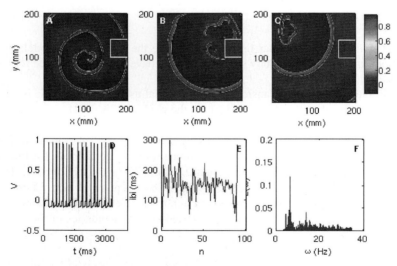

Fig. 4 Spiral turbulence in the Panfilov model with a square obstacle of side 40 mm placed with its lower-left corner at (100 mm, 160 mm). We start with the initial condition of Fig. 2. The spiral wave first gets attached to the obstacle but eventually break up occurs as illustrated in the pseudocolour plots of V in (**A**), (**B**), and (**C**) at 1,100 ms, 1,650 ms, and 2,750 ms, respectively. The local time series for V, the associated inter-beat interval (IBI), and the power spectrum from a representative point in the simulation domain are shown in (**D**), (**E**), and (**F**), respectively; the nonperiodic behaviour of the IBI and the broad-band nature of the power spectrum are characteristic of the underlying spiral turbulence

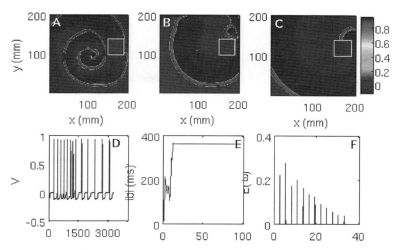

Fig. 5 Pseudocolour plots illustrating the attachment of a spiral wave to a square obstacle of side 40 mm, in the Panfilov model, with its *lower-left corner* is at (100 mm, 150 mm) and the initial condition of Fig. 2. (**A**), (**B**), and (**C**) show pseudocolour plots of V at 1,100 ms, 1,650 ms, and 2,750 ms, respectively. The wave gets attached to the obstacle after about 9 spiral rotations ($\simeq 1,800$ ms). The resulting periodic behviour is reflected in (**D**) the time series of V and (**E**) the plot of the inter-beat interval IBI versus the spike number n: after transients the IBI settles to a constant value of 363 ms; (**F**) the power spectrum has discrete peaks at the fundamental frequency $\omega_f = 2.74$ Hz and its harmonics

figures show, we can obtain ST, RS, or Q states in both these models depending on where the obstacle is placed; pseudocolour plots of V show broken spirals, one rotating spiral, or no spirals in ST, RS, and Q states, respectively; the time series, inter-beat intervals, and power spectra (from a representative point in the simulation domain) are also shown in Figs. 4, 5, 6, 7, 8, 9 for these three states.

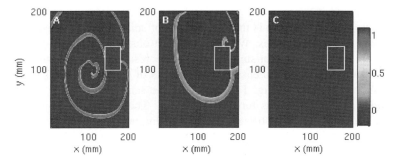

Fig. 6 Pseudocolour plots illustrating how, in the Panfilov model, a spiral wave moves away from the simulation domain if a square obstacle of side 40 mm is placed with its *lower-left corner* at (100 mm, 140 mm) and we use the initial condition of Fig. 2. (**A**), (**B**), and (**C**) show pseudocolour plots of V at 1,100 ms, 1,650 ms and 2,750 ms, respectively

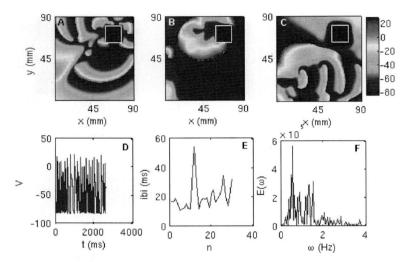

Fig. 7 Spiral turbulence in the presence of an obstacle in the LRI model: We illustrates how the initial condition (Fig. 3A) evolves in the presence of an obstacle of side $l = 18$ mm placed with its *bottom-left corner* at (58.5 mm, 63 mm); spiral turbulence persists in this case; (**A**), (**B**), and (**C**) show pseudocolour plots of V at 200, 600, and 1,000 ms, respectively. (**D**) The local time series from (45 mm, 45 mm); (**E**) the inter-beat interval (IBI) for this time series; and (**F**) the power spectrum for the time series in (**D**) [from 261424 timesteps (i.e., 2614.24 ms)]

Our conclusions are summarized in the stability diagram of Fig. 10. This gives the effect of a square conduction inhomogeneity, with side $l = 40$ mm, in a square simulation domain with side $L = 200$ mm. The colours of the small squares (of side l_p) in Fig. 10 indicate the final state of the system when the position of the bottom-left corner of the obstacle coincides with that of the small square; red, blue, and green denote, respectively, spiral turbulence (ST), a single rotating spiral (RS) anchored at the obstacle, and a quiescent state (Q). The boundaries between ST, RS, and Q appear to be fractal-like as suggested by the enlarged versions of Fig. 10A in Fig. 10B and 10C.

The next issue we address is the dependence of spiral-wave dynamics in these models on additional obstacles. We place the first obstacle A at (x_A, y_A) and the next obstacle B at (x_B, y_B). For simplicity we show results only from our studies of the Panfilov model in two dimensions. The effects of obstacle A are summarized in the stability diagram of Fig. 11A that uses the same colour code as Fig. 10; we consider only 18 positions for the obstacle A. The obstacle B is now kept at (120 mm, 110 mm). If only obstacle B were in this position without obstacle A, spiral break up would have continued. Now we study the combined effects of both these obstacles on the spiral waves.

We note first that, in most cases, the effects of one of the obstacles dominate spiral-wave dynamics. For example, when obstacle A alone is placed at $(x_A = 120$ mm, $y_A = 60$ mm) the spiral wave is anchored to it; the addition of a second

The Mathematical Modelling of Inhomogeneities in Ventricular Tissue

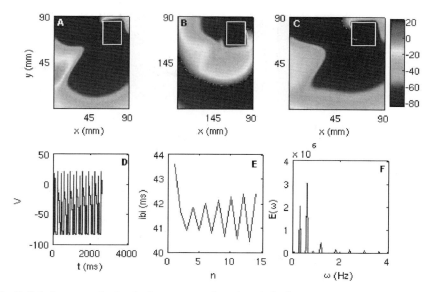

Fig. 8 Spiral-wave anchoring in the presence of an obstacle in the LRI model: We illustrate how the initial condition (Fig. 3A) evolves in the presence of an obstacle of side $l = 18$ mm with its *bottom-left corner* at (58.5 mm, 63 mm) to a spiral wave anchored to the obstacle. (**A**), (**B**), and (**C**) show pseudocolour plots of V at 200, 600, and 1,000 ms, respectively. (**D**) The local time series from (45 mm, 45 mm); (**E**) the inter-beat interval (IBI) versus the spike number n for this time series; and (**F**) the power spectrum for the time series in (**D**) [from 261,424 timesteps (i.e., 2614.24 ms)]

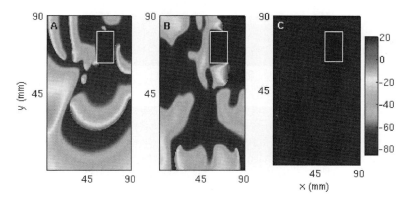

Fig. 9 A spiral-wave moving away from the simulation domain in the presence of an obstacle in the LRI model: We illustrate how the initial condition (Fig. 3A) evolves in the presence of an obstacle of side $l = 18$ mm with its *bottom-left corner* at (54 mm, 63 mm). (**A**), (**B**), and (**C**) show pseudocolour plots of V after 200 ms, 600 ms, and 1,000 ms, respectively

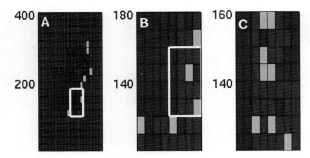

Fig. 10 Panfilov-model stability diagram: The effect of a square conduction inhomogeneity with side $l = 40$ mm in a square simulation domain with side $L = 200$ mm. In this diagram the colours of the *small squares* (of side l_p) indicate the final state of the system when the position of the *bottom-left corner* of the obstacle coincides with that of the *small square*; the colours *red*, *blue*, and *green* denote, respectively, spiral turbulence (ST), a single rotating spiral (RS) anchored at the obstacle, and a quiescent state (Q). (**A**) for $l_p = 10$ mm. We expose the fractal-like structure of the boundaries between ST, RS, and Q by zooming in on small subdomains encompassing parts of these interfaces (white boundaries in (**A**) and (**B**)) with (**B**) $l_p = 5$ mm, (**C**) and $l_p = 2.5$ mm

obstacle does not change this; similarly, when obstacle A alone is at ($x_A = 90$ mm, $y_A = 50$ mm), the spiral moves away from the simulation domain; this too is not affected by the presence of the second obstacle. However, if obstacle A is at ($x_A = 110$ mm, $y_A = 50$ mm), the spiral wave gets anchored to it; now if the second

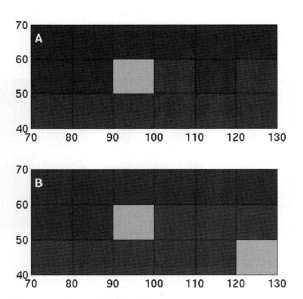

Fig. 11 (**A**) Stability diagram for the Panfilov model ($L = 200$ mm and $l_p = 10$ mm) with a *square obstacle* **A**, of side $l = 40$ mm. The colour code is the same as the one used in Fig. 10. (**B**) The analogue of (**A**) but with two *square obstacles* **A** and **B** of the same size. Obstacle **B** is placed at (120 mm, 110 mm). The colour of a *square* shows the final state of the system when the *left bottom corner* of obstacle **A** coincides with that of the *square*

obstacle B is placed at ($x_B = 120$ mm, $y_B = 110$ mm), the spiral waves breaks up. The most interesting case occurs if obstacle A is at (120 mm, 40 mm) and obstacle B at (120 mm, 110 mm); in the absence of the other, each one of these obstacles causes the spiral waves to break up; but together they cause these waves to move away from the medium!

4 Conclusions

We have given an overview of studies of the effects of conduction inhomogeneities on spiral-wave dynamics in models for ventricular tissue. In particular, we have presented our extensive and systematic work on this problem in the Panfilov and LRI models. The most remarkable qualitative conclusion of our study is that spiral-wave dynamics in these models depends very sensitively on the positions of conduction inhomogeneities. All possible behaviours, ST, RS, and Q are possible as shown in the stability diagrams of Figs. 9 and 10; and the boundaries between these appear to be fractal like. As we have argued in Ref. [22], this must arise because of an underlying fractal basin boundary between the domains of attraction of the ST, RS, and Q states in our high-dimensional dynamical systems, the Panfilov and LRI models (strictly speaking these are infinite-dimensional dynamical systems since they are PDEs). Normally one goes from one domain of attraction to another in a dynamical system by changing the initial condition. Here we do not change the initial condition but we modify the dynamical system slightly by changing the position of the obstacle. Our studies show that the final state of the system depends very sensitively on this change.

Our study has obvious and important implications for cardiac arrhythmias. It suggests that such arrhythmias must depend sensitively on the positions of conduction inhomogeneities in cardiac tissue. We believe this qualitative insight will help in the development of low-amplitude defibrillation schemes that can work even in the presence of inhomogeneities in cardiac tissue.

Acknowledgments We thank DST, UGC, and CSIR (India) for support, and SERC (IISc) for computational resources.

Appendix: The Luo Rudy I Model

In the Luo-Rudy I (LR I) model there are six components of the ionic current, which are formulated mathematically in terms of Hodgkin-Huxley-type equations. The partial differential equation for the transmembrane potential V is

$$\frac{\partial V}{dt} = \frac{I_{LR}}{C} = D\nabla^2 V. \qquad (2)$$

Here I_{LR} is the instantaneous, total ionic-current density. The subscript *LR* denotes that we use the formulation of the total ionic current described by the Luo-Rudy Phase I (LR1) model [25]. In the LR1 model, $I_{LR} = I_{Na}+I_{si}+I_K+I_{K1}+I_{Kp}+I_b$, with the current densities I_{Na} (fast inward Na$^+$), I_{si} (slow inward), I_K (slow outward time-dependent K$^+$), I_{K1} (time-independent K$^+$), I_{Kp} (plateau K$^+$), I_b (total background), given by:

$$I_{Na} = G_{Na}m^2hj(V - E_{Na})$$
$$I_{si} = G_{si}df(V - E_{si})$$
$$I_K = G_k x x_i (V - E_K)$$
$$I_{K_1} = G_K K_{1\infty}(V - E_{K_1})$$
$$I_{Kp} = G_{Kp} K_p (V - E_{Kp})$$
$$I_b = 0.03921(V + 59.87)$$

and $K_{1\infty}$ is the steady-state value of the gating variable K_1. All current densities are in units of $\mu A/cm^2$, voltages are in mV, and G_ξ and E_ξ are, respectively, the ion-channel conductance and reversal potential for the channel ξ. The ionic currents are determined by the time-dependent ion-channel gating variables $h, j, m, d, f, x, x_i, K_p$ and K_1 generically denoted by ξ, which follow ordinary differential equations of the type

$$\frac{d\xi}{dt} = \frac{\xi_\infty - \xi}{\tau_\xi}$$

where $\xi_\infty = \alpha_\xi/(\alpha_\xi + \beta_\xi)$ is the steady-state value of ξ and $\tau_\xi = \frac{1}{\alpha_\xi + \beta_\xi}$ is its time constant. The voltage-dependent rate constants, α_ξ and β_ξ, are given by the following empirical equations:

$$\alpha_h = 0, \text{ if } V \geq -40\,mV,$$
$$= 0.135 \exp[-0.147(V + 80)],$$

otherwise;

$$\beta_h = \frac{1}{0.13(1 + \exp[-0.09(V + 10.66)])}, \text{ if } V \geq -40\,mV$$
$$= 3.56 \exp[0.079\,V] + 3.1 \times 10^{-5} \exp[0.35\,V],$$

otherwise;

$$\alpha_j = 0, \text{ if } V \geq -40\,mV$$
$$= \left[\frac{(\exp[0.2444V] + 2.732 \times 10^{-10} \exp[-0.04391V])}{-7.865 \times 10^{-6} \{1 + \exp[0.311(V + 79.23)]\}} \right](V + 37.78),$$

otherwise;

$$\beta_j = \frac{0.3 \exp[-2.535 \times 10^{-7} V]}{1 + \exp[-0.1(V + 32)]}, \text{ if } V \geq -40 \, mV$$

$$= \frac{0.1212 \exp[-0.01052 V]}{1 + \exp[-0.1378(V + 40.14)]},$$

otherwise;

$$\alpha_m = \frac{0.32(V + 47.13)}{1 - \exp[-0.1(V + 47.13)]};$$

$$\beta_m = 0.08 \exp[-0.0909 V];$$

$$\alpha_d = \frac{0.095 \exp[-0.01(V - 5)]}{1 + \exp[-0.072(V - 5)]};$$

$$\beta_d = \frac{0.07 \exp[-0.017(V + 44)]}{1 + \exp[0.05(V + 44)]};$$

$$\alpha_f = \frac{0.012 \exp[-0.008(V + 28)]}{1 + \exp[0.15(V + 28)]};$$

$$\beta_f = \frac{0.0065 \exp[-0.02(V + 30)]}{1 + \exp[-0.2(V + 30)]};$$

$$\alpha_x = \frac{0.0005 \exp[0.083(V + 50)]}{1 + \exp[0.057(V + 50)]};$$

$$\beta_x = \frac{0.0013 \exp[-0.06(V + 20)]}{1 + \exp[-0.04(V + 20)]};$$

$$\alpha_{K1} = \frac{1.02}{1 + \exp[0.2385(V - E_{K1} - 59.215)]};$$

$$\beta_{K1} = \left[\frac{0.49124 \exp[0.08032(V - E_{K1} + 5.476)]}{1 + \exp[-0.5143(V - E_{K1} + 4.753)]} \right. $$
$$\left. + \exp[0.06175(V - E_{K1} - 594.31)] \right].$$

The gating variables x_i and K_p are given by

$$x_i = \frac{2.837 \exp 0.04(V + 77) - 1}{(V + 77) \exp 0.04(V + 35)}, \text{ if } V > -100 \, mV,$$

$$= 1, \text{ otherwise};$$

$$K_p = \frac{1}{1 + \exp[0.1672(7.488 - V)]}.$$

The values of the channel conductances G_{Na}, G_{si}, G_K, G_{K1}, and G_{Kp} are 23, 0.07, 0.705, 0.6047 and 0.0183 mS/cm^2, respectively. The reversal potentials are $E_{Na} = 54.4$ mV, $E_K = -77$ mV, $E_{K1} = E_{Kp} = -87.26$ mV, $E_b = -59.87$ mV, and $E_{si} = 7.7 - 13.0287 \ln Ca$, where Ca is the calcium ionic concentration satisfying

$$\frac{dCa}{dt} = -10^{-4} I_{si} + 0.07(10^{-4} - Ca).$$

The times t and τ_ξ are in ms; the rate constants α_ξ and β_ξ are in ms^{-1}.

References

1. J. R. Hampton, *The ECG Made Easy*, Churchill Livingstone (2003).
2. M. C. Cross and P. C. Hohenberg, *Pattern formation outside equilibrium*, Rev. Mod. Phys., **65**, 851 (1993).
3. A. T. Winfree, Evolving perspectives during 12 years of electrical turbulence, Chaos, **8**, 1 (1998).
4. A. T. Winfree, *When Time Breaks Down*, Princeton University Press (1987).
5. J. Jalife, R. A. Gray, G. E. Morley, and J. M. Davidenko, *Spiral breakup as a model of ventricular fibrillation*, Chaos, **8**, 57 (1998).
6. R. Pandit, A. Pande, S. Sinha, and A. Sen, *Spiral turbulence and spatiotemporal chaos: characterization and control in two excitable media*, Physica A, **306**, 211 (2002).
7. T. Ikeda, M. Yashima, T. Uchida, D. Hough, M. C. Fishbein, W. J. Mandel, P.-S. Chen, and H. S. Karagueuzian, *Attachment of meandering reentrant wave fronts to anatomic obstacles in the atrium: role of obstacle size*, Circ. Res., **81**, 753 (1997).
8. J. M. Davidenko, A. V. Pertsov, R. Salomonsz, W. T. Baxter, and J. Jalife, *Stationary and drifting spiral waves of excitation in isolated cardiac muscle*, Nature, **355**, 349 (1929).
9. M. Valderrabano, P. S. Chen, and S. F. Lin, *Spatial distribution of phase singularities in ventricular fibrillation*, Circulation, **108**, 354 (2003).
10. M. Valderrabano, M.-H. Lee, T. Ohara, A. C. Lai, M. C. Fishbein, S.-F. Lin, H. S. Karagueuzian, and P. S. Chen, *Dynamics of intramural and transmural reentry during ventricular fibrillation in isolated swine ventricles*, Circ. Res., **88**, 839–848 (2001).
11. T.-J. Wu, J. J. Ong, C. Hwang, J. J. Lee, M. C. Fishbein, L. Czer, A. Trento, C. Blanche, R. M. Kass, W. J. Mandel, et al. *Characteristics of wave fronts during ventricular fibrillation in human hearts with dilatedcardiomyopathy: role of increased fibrosis in the generation of reentry*, J. Am. Coll. Card., **32**, 187 (1998).
12. T. Ohara, K. Ohara, J.-M. Cao, M.-H. Lee, M. C. Fishbein, W. J. Mandel, P.-S. Chen, and H. S. Karagueuzian, *Increased wave break during ventricular fibrillation in the epicardial border zone of hearts with healed myocardial infarction*, Circulation, **103**, 1465–1472 (2001).
13. S.-M. Hwang, T. Y. Kim, and K. J. Lee, *Complex-periodic spiral waves in confluent cardiac cell cultures induced by localized inhomogeneities*, Proc. Natl. Acad. Sci. USA, **102**, 10363 (2005).
14. F. Xie, F. Qu, and A. Garfinkel, *Dynamics of reentry around a circular obstacle in cardiac tissue*, Phys. Rev. E, **58**, 6355 (1998).
15. A. M. Pertsov, J. M. Davidenko, R. Salomonsz, W. T. Baxter, and J. Jalife, *Spiral waves of excitation underie reentrant activityisolated cardiac muscle*, Circ. Res., **72**, 631 (1993).

16. M. Valderrabano, Y.-H. Kim, M. Yasima, T.-J. Wu, H. S. Karagueuzian, and P.-S. Chen, *Obstacle-induced transition from ventricular fibrillation to tachycardia in isolated swine right ventricles*, J. Am. Coll. Cardiol., **36**, 2000 (2000).
17. Y. H. Kim, F. Xie, M. Yashima, T.-J. Wu, M. Valderrabano, M.-H. Lee, T. Ohara, O. Voroshilovsky, R. N. Doshi, M. C. Fishbein, et al. *Role of papillary muscle in the generation and maintenance of reentry during ventricular tachycardia and fibrillation in isolated swine right ventricle*, Circulation, **100**, 1450 (1999).
18. J. M. Starobin and C. F. Starmer, *Boundary-layer analysis of waves propagating in an excitable medium: medium conditions for wave-front-obstacle separation*, Phys. Rev. E, **54**, 430 (1996).
19. S. Takagi, A. Pumir, D. Pazo, I. Efimov, V. Nikolski, and V. Krinsky, *Unpinning and removal of a rotating wave in cardiac tissue*, Phys. Rev. Lett., **93**, 058101 (2004).
20. A. V. Panfilov, *Spiral breakup in an array of coupled cells: the role of the intracellular conductance*, Phys. Rev. Lett., **88**, 118101 (2002).
21. K. H. W. J. ten Tusscher and A.V. Panfilov, *Influence of nonexcitable cells on spiral breakup in two-dimensional and three-dimensional excitable media*, Phys. Rev. E, **68**, 062902 (2003).
22. T. K. Shajahan, S. Sinha, and R. Pandit, *Spiral-wave dynamics depend sensitively on inhomogeneities in mathematical models of ventricular tissue*, Phys. Rev. E, **75**, 011929 (2007).
23. A. V. Panfilov, *Spiral breakup as a model of ventricular fibrillation*, Chaos, **8**, 57 (1998).
24. A. V. Panfilov and P. Hogeweg, Phys. Lett. A, **176**, 295 (1993).
25. C. H. Luo and Y. Rudy, *A model of the ventricular cardiac action potential: depolarization, repolarization, and their interaction*, Circ. Res., **68**, 1501 (1991).
26. T. K. Shajahan, S. Sinha, and R. Pandit, *Ventricular fibrillation in a simple excitable medium model of cardiac tissue*, Int. J. Mod. Phys. B, **17**, 5645 (2003).
27. T. K. Shajahan, S. Sinha, and R. Pandit, *Spatiotemporal chaos and spiral turbulence in models of cardiac arrhythmias*, Proc. Indian Natl. Sci. Acad., **71** A, 4757 (2005).
28. A. Pande and R. Pandit, *Spatiotemporal chaos and nonequilibrim transitions in a model excitable medium*, Phys. Rev. E, **61**, 6448 (2000).
29. S. Sinha, A. Pande, and R. Pandit, *Defibrillation via the elimination of spiral turbulence in a model for ventricular fibrillation*, Phys. Rev. Lett., **86**, 3678 (2001).

Controlling Spiral Turbulence in Simulated Cardiac Tissue by Low-Amplitude Traveling Wave Stimulation

Sitabhra Sinha and S. Sridhar

Abstract Ventricular fibrillation (VF), a class of cardiac arrhythmias that is often fatal, is associated with the breakdown of spatially coherent activity in heart tissue. Modeling studies have linked VF with the onset of spatiotemporal chaos in the electrophysiological activity of the heart, through the creation and subsequent breakup of spiral waves. Conventionally, defibrillation is carried out by applying large electrical shocks to the heart which has harmful effects both in the short and long terms. Using nonlinear dynamics techniques, several low-amplitude control methods for VF have been suggested. However, all of them suffer from the problem of either having to use high power (applied over the entire system) or extremely high frequencies (which are unstable and may spontaneously give rise to further VF episodes). In this paper we propose a spatially extended but non-global scheme for controlling VF in simulated cardiac tissue. The method involves successive activation of electrodes arranged in a square array, such that, a wave of control stimulation is seen to propagate through the system. Our simulations involving both simple and realistic models of ventricular tissue show that spatiotemporal chaotic activity associated with VF can be terminated using low amplitude control.

Keywords Cardiac fibrillation · spatiotemporal chaos · spiral turbulence · chaos control · excitable media

1 Introduction

Cardiac arrhythmias, which are disturbances in the natural rhythmic activity of the heart, are a significant contributor to cardiac diseases, the leading cause of death in the industrialized world [1]. There are several types of arrhythmia, among which ventricular fibrillation (VF) is the one most likely to be fatal. Fibrillation is described as failure of coherent activity in cardiac muscle, as a result of which the heart stops pumping blood. If not treated immediately, the patient dies within a few minutes.

S. Sinha (✉)
The Institute of Mathematical Sciences, CIT Campus, Taramani, Chennai 600113, India
e-mail: sitabhra@imsc.res.in

Treatment of fibrillation by means of pharmaceutical drugs has so far proved relatively ineffective, and, till date, the usual treatment involves applying an electrical shock to the heart in order to "reset" its activity (electrical defibrillation). Conventional methods of defibrillation involve applying large currents either externally ($\simeq 1$ A) or internally to the heart, for a short duration (a few ms). The method works on the principle that these large amplitude shocks excite the entire heart tissue in a synchronized fashion, thereby terminating all irregular electrical activity, allowing the sinus node, which is the natural cardiac pacemaker, to take over and ensure normal rhythmic behavior. However, using such large currents can damage cardiac tissue, resulting in the creation of scar tissue which can potentially act as substrate for future arrhythmias. Further, these shocks can be extremely painful, often making the patient unconscious. These reasons have led to research in devising alternative defibrillation methods using very low-amplitude electrical stimuli. In recent years, physicists have used the principles of nonlinear dynamics, in particular, techniques of chaos control, to suggest efficient schemes for terminating life-threatening arrhythmias. In this paper, we have proposed a method for controlling VF which uses extremely low amplitude control signals applied for a brief duration over an array of electrodes. In the following subsections, we introduce our modeling methodology involving excitable media, in particular, the generation of spatiotemporal chaos resembling VF in such media and give a brief review of the various control schemes that have been suggested so far. In the next section we describe in detail our control algorithm, followed by a section on our results. Certain notable aspects of the control scheme are explored in the section containing discussions. We conclude with a brief consideration of the applicability of our method for defibrillation in a clinical setting.

1.1 Excitable Media

In order to model cardiac dynamics, one can focus exclusively on the electrical excitation behavior of cardiac myocytes. The mechanical activity of these cells which results in the pumping action of the heart is essentially a consequence of their electrical activity. Biological cells that can be electrically excited are often said to possess the property of *excitability*. As excitation can spread from a cell to its neighbors through diffusion via gap junctions, cardiac tissue can be accurately described by a class of models collectively referred to as *excitable media* [2].

An excitable media is defined by the following properties:

- The dynamical behavior is characterized by a stable *resting* state and a metastable *excited* state, the system moving from the former to the latter if it is perturbed by a stimulus that exceeds a specified *threshold*, V_{th}. Upon being excited, the system gradually recovers and returns to the resting state describing a characteristic time-evolution profile referred to as an *action potential*. For cardiac myocytes, the state of each cell is described by the potential difference across the cell membrane, V, with the resting state characterized by the resting membrane potential $V_{\text{rest}} \simeq 85$ mV, and the threshold for stimulation by $V_{\text{th}} \simeq 60$ mV.

- After being excited once, a cell cannot be excited again even by a supra-threshold stimulation for a duration which is referred to as the *refractory period*. For cardiac cells, this period can be divided into the *absolute* and *relative* refractory periods. In the former, no stimulus (regardless of amplitude) can elicit an action potential, while in the latter, a stimulus having amplitude significantly higher than the threshold V_{th} can succeed in exciting the cell, although the resulting action potential has a much shorter duration than normal.

The simplest set of equations which exhibit the above features and is therefore used to describe excitable media is the generic Fitzhugh-Nagumo (FHN) system:

$$\partial e/\partial t = D\nabla^2 e + e(1-e)(e-c) - g,$$
$$\partial g/\partial t = \varepsilon(ke - g),$$

where, the dynamics of a cell is described by using a fast (or excitation) variable e, representing the trans-membrane potential and a slow (recovery) variable g, that represents the effective membrane conductance. The latter is an aggregated variable representing all the complexities of various cellular ion channels occurring in reality. Further, D is the diffusion constant governing the propagation speed of excitation across cells, and ε is the ratio of time scales over which the two variables evolve, which controls the duration of the refractory period. The resting state corresponds to $e = 0$, while the excited state is $e = 1$, with the threshold given by $e = c$.

Panfilov [3, 4] has proposed a model of cardiac excitation based on the FHN equations which implements the two types of refractory period seen in cardiac cells, viz., absolute and refractory. It also uses a piecewise linear function, instead of the cubic nonlinearity of the standard FHN equations. The model is defined as

$$\partial e/\partial t = D\nabla^2 e - f(e) - g,$$
$$\partial g/\partial t = \varepsilon(e,g)(ke - g).$$

As in the FHN system, the ratio of time scales for the two variables is given by ε. However, its value depends on the position of the system at any given time in the (e,g) phase space: $\varepsilon(e,g) = \varepsilon_1$ for $e < e_2$, $\varepsilon(e,g) = \varepsilon_2$, for $e > e_2$ and $\varepsilon(e,g) = \varepsilon_3$ for $e < e_1$ and $g < g_1$ with $e_1 = 0.0026$, $e_2 = 0.837$, $g_1 = 1.8$, $\varepsilon_1 = 1/75$, $\varepsilon_2 = 1.0$ and $\varepsilon_3 = 0.3$. The piecewise linear function, $f(e)$, is defined as: $f(e) = C_1 e$, for $e < e_1$, $f(e) = -C_2 e + a$, for $e_1 \leq e \leq e_2$, and $f(e) = C_3(e-1)$, for $e > e_2$, with $C_1 = 20$, $C_2 = 3$, $C_3 = 15$, $a = 0.06$ and $k = 3$. In principle, D is a tensor quantity; however, in our modeling D is taken to be 1 for simplicity.

As mentioned above, the Panfilov model does not explicitly consider realistic biological features, such as the different types of ion channels in the cell membrane. A class of more realistic models has been proposed which explicitly takes into account the dynamics of such channels. The general form of such models follows the Hodgkin-Huxley formulation [5] with the time-evolution of the trans-membrane potential, V, described by the following equation:

$$\partial V/\partial t = -(I_{\text{ion}}/C_{\text{m}}) + D\nabla^2 V.$$

Here, C_{m} is the membrane capacitance density, D is the diffusion constant and I_{ion} is the instantaneous total ionic current density. The various realistic models for cardiac excitation differ in the formulation of I_{ion}. In the Luo-Rudy I model [6] of guinea pig ventricular myocyte cells, this current density is assumed to be composed of six different ionic current densities,

$$I_{\text{ion}} = I_{\text{Na}} + I_{\text{si}} + I_{\text{K}} + I_{\text{K1}} + I_{\text{Kp}} + I_{\text{b}}.$$

Here, I_{Na} is the fast sodium current, I_{si} is the slow inward current, I_{K} is the time-dependent potassium current, I_{K1} is the time-independent potassium current, I_{Kp} is the plateau potassium current and I_{b} is the background leakage current. These currents are in turn determined by several time-dependent ion-channel gating variables whose time-evolution is governed by ordinary differential equations of the form,

$$d\xi/dt = (\xi_{\infty} - \xi)/T_{\xi}.$$

Here, ξ_{∞} is the steady state value of ξ and T_{ξ} is the time constant. These parameters are complicated functions of V and are obtained by fitting experimental data with exponential functions. In the work reported in this paper, we have carried out our simulations with both the Panfilov as well as Luo-Rudy I model.

1.2 Spatiotemporal Chaos in Excitable Media

In excitable media, wave-fronts mutually annihilate on collision, as behind each front is an in-excitable region which is in the refractory state. Therefore, analogous to colliding fire-fronts, the two wave fronts cannot penetrate each other because the excitation cannot propagate through the refractory region immediately behind each of these fronts. This property gives rise to a variety of spatial patterns when multiple excitation wave-fronts interact. In particular, spiral waves (in two-dimensional media) and scroll waves (in three dimensions) are formed due to time evolution of free (or broken) ends of excitation fronts, which might occur spontaneously through interactions with in-excitable regions. In our simulations, spiral waves are initiated by allowing a partial planar wave-front (extending over only a fraction of the system's spatial extent) to evolve in time, such that its free end gradually curls into a spiral wave. In a three dimensional media, a cylindrical wavefront is allowed to propagate through the medium until it is broken such that the free interface gradually curls inward. The resulting scroll wave can be thought of as a sequence of spiral waves stacked on top of one another with their tips joined together forming the scroll wave filament. Once formed, such waveforms are self sustaining sources of high frequency excitation. Depending on the system parameters, these waves can become unstable and break up to form smaller spiral (or scroll) wavelets, eventually resulting in a spiral turbulent state. This state is characterized by spatiotemporally irregular

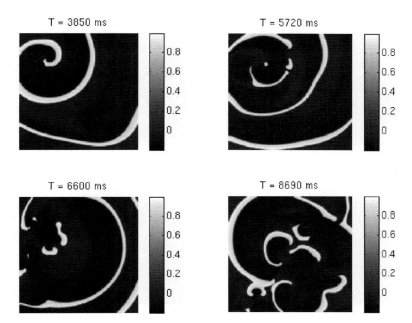

Fig. 1 The onset of spatiotemporal chaos shown in a pseudo-grayscale plot of the e-variable in a 2-dimensional Panfilov model system having linear dimension $L = 12.8$ cm. The initial condition is a broken plane wave-front which is allowed to curl around into a spiral wave. Meandering of the spiral wave causes it to break up into many smaller spirals leading to a spiral turbulent state

excitation over the media and complete absence of any coherent activity. The high-dimensional chaotic nature of this state is demonstrated from the corresponding Lyapunov spectrum and Kaplan-Yorke dimension [7]. Experiments by Witkowski et al. [8] and Gray et al. [9] have identified the turbulent cardiac activity during VF to be associated with the formation and subsequent breakup of electrophysiological structures (*rotors*) which emit spiral waves. The Panfilov model is one of the simplest excitable media models which display such breakup of spiral waves leading to spatiotemporal chaotic activity (Fig. 1). Similar structures are also seen for the more realistic Luo-Rudy I model (Fig. 2).

1.3 A Brief Review of Control Methods for Spatiotemporal Chaos in Excitable Media

Control of spatiotemporal chaos in an excitable medium have certain special features. For example, the existence of an excitation threshold would imply that the response of the medium to a control signal is not proportional to the strength of the signal. Hence, techniques such as linear proportional feedback which are normally used for chaos control in other situations are not applicable here. Further, as VF is

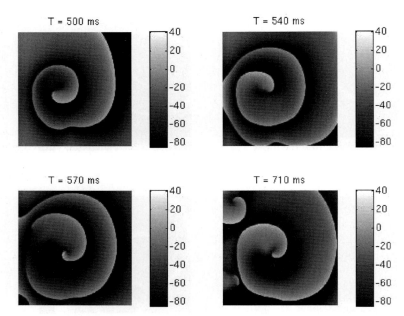

Fig. 2 The onset of spatiotemporal chaos shown in a pseudo-grayscale plot of the trans-membrane potential V in a 2-dimensional Luo-Rudy I model system having linear dimension $L = 9$ cm. The initial condition is same as Fig. 1. The colorbar indicates values of V in mV

fatal within minutes, the aim of any control scheme would be to remove all activity from the system as quickly as possible, so that the natural pacemaker of the heart can take over again. Various methods have been proposed to terminate spatiotemporal chaos in excitable media using low amplitude signals. Depending on the method in which the control signal is applied to the system, they can be broadly classified into three groups, viz. global control, local control and non-global spatially extended control.

1.3.1 Global Control

Global control involves the application of the control signal at all points in the media, although the signal amplitude may vary from region to region. While global control is hard to implement experimentally, there have been several schemes proposed that have worked well in numerical simulations. The general scheme involves applying an external control stimulus A for a duration T. If A is a small additive perturbation in the time evolution equation for the trans-membrane potential (e.g., the variable e in the Panfilov model), it results in an effective reduction of the excitation threshold. This, in turn, promotes the simultaneous excitation of different regions. The control stimulus can in general be temporally periodic, consisting of a sequence of pulses. The signal amplitude can also be periodic in space, as well as time, as has been used by Wang et al. [10].

Instead of perturbing the membrane potential, Osipov and Collins [11] have used additive perturbation to the recovery variable (g in the Panfilov model) representing effective membrane conductance. This results in different wave-front and wave-back velocities, which destabilizes the traveling wave and eventually leads to elimination of all activity. Gray [12] has found that using long duration pulses affecting both membrane potential and effective ionic conductance, as opposed to short duration pulses which affects only the membrane potential, is a more efficient control method as it uses lower power.

Other control schemes involve applying perturbation directly to model parameters such as excitation threshold, which may not be experimentally accessible. For example, Alonso et al. [13, 14] suggested a periodic perturbation of the excitation threshold to control spiral turbulence in three dimensional excitable media. The result of the periodic forcing depends on the difference between the control frequency and spiral rotation frequency. Rapid forcing increases the positive tension of the spiral filaments, causing them to shrink and collapse, leading to the elimination of spatiotemporal chaos. Spatially uncorrelated Gaussian noise have also been used to suppress spiral wave turbulence [15].

1.3.2 Local Control

Global schemes are very difficult to implement in clinical situations and would also involve high power consumption. These problems can be solved by local control schemes, which involve applying the control signal at only a small localized region of the spatially extended system. Almost all such methods use the idea of overdrive pacing to eliminate spatiotemporal chaos. A series of waves with frequency higher than any of the existing excitations in the turbulent medium is generated with the aim of sweeping away the chaotic activity to the non-conducting boundary of the system where they are absorbed. Eventually, the medium is completely occupied by waves generated through local pacing and the system returns to the quiescent state when the control is switched off. The time to achieve control is inversely related to the difference between the overdrive pacing frequency and the dominant frequencies of the chaotic activity.

Stamp et al. [16] used a series of pulses of different waveform shapes to terminate spiral turbulence, but only met with limited success. Periodic stimulations were used by Zhang et al. [17] to successfully eliminate spatiotemporal chaos in certain models. The fundamental mechanism of this scheme involves periodic alternation between positive and negative stimulation. Reference [18] explores this biphasic control method in detail and suggests efficient waveform shapes for rapidly terminating spiral turbulence. The negative pulses applied at the control point leads to shortening of the recovery period around the region of stimulation, thereby allowing generation of very high frequency waves which drive away the chaotic activity. Other schemes have proposed using local perturbation of experimentally inaccessible parameters like the excitation threshold [19]. These also involve generating extremely high frequency waves.

While a local scheme would be easy to implement, all the methods proposed involve generating very high frequency control waves. These waves become unstable as they propagate away from the stimulation point and can breakup, thereby reinitiating spiral waves in the medium. Also, it is known that in the presence of inhomogeneities, rapid pacing can lead to spatiotemporal chaos, thus making local control methods unsuitable for actual implementation [20].

1.3.3 Non-Global Spatially Extended Control

When the control signal is applied at a large number of spatially distant regions, but not at all points of the medium, we obtain an intermediate form of control method, referred to as a non-global spatially extended scheme. Sinha et al. [21] have developed a control method involving supra-threshold stimulation applied along a grid of line-electrodes. The fundamental idea behind this scheme is that the duration of spatiotemporal chaos in excitable media is an exponentially increasing function of the system size. By exciting thin strips of medium using the line electrodes, the system is divided into electrically disconnected regions separated by boundaries of refractory cells. Each of these regions is too small to sustain chaotic activity for any significant period of time. As a result, the spiral waves in each such region are absorbed by their boundaries (which are in the refractory state, and hence, non-conducting) so that all activity is terminated within a short interval. While this method is effective even in the presence of inhomogeneities, its implementation is complicated by the fact that line electrodes are difficult to implant in a clinical setting.

Clinical experiments done so far in controlling fibrillation using low energy defibrillators have not been very successful. Gauthier et al. [22], used a closed loop feedback control protocol where the control perturbations are applied at a single spatial location to sheep atria *in vivo*. Their control scheme was found to be not effective in regularizing the cardiac dynamics. Mitchell et al. [23] found that atrial pacing is ineffective for treating persistent atrial fibrillation, although it does terminate atrial tachcycardia, a simpler type of arrhythmia usually associated with a single spiral wave. Note that, both these methods involved pacing from a localized electrode. As mentioned above, spatially extended chaotic activity is difficult to control using local stimulation, especially in the presence of inhomogeneities, as would be the case in a real experimental or clinical situation.

In this paper, we use an array of control points placed at regular intervals to eliminate spiral turbulence in excitable media. While this is an instance of a non-global spatially extended control method, the use of localized point electrodes (rather than line electrodes, as in Ref. [21]) should make it easier to implement in an experimental or clinical setting. Moreover, we show it to be robust in the presence of inhomogeneities, unlike the local pacing schemes described above.

2 Array Control Scheme and Simulation Details

Our scheme for controlling spatiotemporal chaos in excitable media uses a traveling wave for stimulating different regions of the system ensuring that all high frequency sources of excitation in the turbulent regions are engaged by the control signal. The control scheme is implemented by introducing a square array of control electrodes spaced distance d apart, each of which are in contact with the simulated cardiac tissue. On initiation of control, the electrode at one corner of the system is switched on to apply a low-amplitude stimulus for a brief duration ($\simeq$ a few ms) to the corresponding portion of the excitable media, followed by successive activation of neighboring electrodes such that a wave of control is seen to travel radially across the simulated tissue away from the original site of stimulation. In certain situations we needed more than one such "wave" to terminate spatiotemporal chaos. The propagation velocity of the control wave, v_{control}, is constant for a given realization. However, we have varied this velocity over a large range by carrying out a number of realizations.

We have performed simulations in both 2- and 3-dimensional domains, with systems as large as 400×400 and $128 \times 128 \times 8$ cells, respectively. We solve both Panfilov and Luo-Rudy I systems using a forward-Euler integration scheme. For the Panfilov model simulations, we discretize the system on a grid of points in space with the interval $dx = 0.5$ dimensionless units. The standard five- and seven-point stencils are used to represent the Laplacian in 2- and 3-dimensional simulations, respectively. The time step is chosen to be $dt = 0.022$ dimensionless units. The dimensioned time and space units are defined to be 5 ms and 1 mm, respectively. The boundary conditions are chosen such that all boundaries of the domain are non-conducting. The dimensioned value of the diffusion constant is $2 \, \text{cm}^2/\text{s}$. For the Luo-Rudy I simulations, we use a time step of $dt = 0.01$ ms and space step of $dx = 0.0225$ cm. The boundary conditions are same as in the Panfilov model simulations. The diffusion constant is taken to be $0.001 \, \text{cm}^2/\text{s}$, such that the conduction speed of excitation waves matches the value seen in real cardiac tissue.

3 Results

Our proposed control scheme is successful in controlling spatiotemporal chaos in both 2-dimensional and 3-dimensional systems. Let us first consider the case of a 2-dimensional system where the dynamics of the cells is described by the Panfilov model. Initially, we consider the limit when the control array spacing $d \to 0$. As shown in Fig. 3, the control scheme is successful in removing all activity within 350 ms after initiation of control. The wave of control stimuli starts from the top left corner of the system and moves across the system with a constant propagation velocity, sweeping away all activity to the boundaries in the process. When the control wave moves out of the system, all activity is seen to be terminated. We observe the existence of a critical amplitude A_c, which is the lowest amplitude

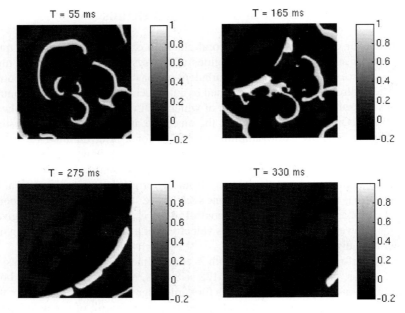

Fig. 3 Spatiotemporal chaos control in a 2-dimensional Panfilov model of linear dimension $L = 12.8$ cm with the control array spacing, $d \to 0$. A stimulus of strength $A = 1.5$ is applied at each control point for an average duration of 15 ms. The control was initiated at $T = 0$ ms

of the control stimulus that succeeds in removing all activity. For control stimuli below this critical value, even several applications of the control do not remove all activity. Complete termination is a necessary condition for successful defibrillation, as any remnant activity can result in re-initiation of turbulent excitation. For the 2-dimensional Panfilov model, the critical amplitude was observed to be $A_c = 1.3$ dimensionless units for $d \to 0$. Next, we verified the efficacy of our control scheme for a finite value of control array spacing d in our simulations. The control scheme is found to work even for $d > 0$, although the critical amplitude increases with d, until the array spacing exceeds the width of the refractory region behind the excitation wave-front.

We next confirm that the control scheme is model independent by using it to eliminate spatiotemporal chaos in the 2-dimensional Luo-Rudy I system. As previously noted, this is a more realistic description of cardiac tissue than the Panfilov model. Upon using a stimulus current $A = 65\,\mu\text{A}/\text{cm}^2$, we find that all activity is eliminated from a system of area $L^2 = 81\,\text{cm}^2$ when the array spacing $d = 0.225$ cm (Fig. 4). With increased spacing between control points, the strength of the stimulus current required for eliminating all activity increases. For example, for $d = 0.27$ cm, a control stimulus of $A = 100\,\mu\text{A}/\text{cm}^2$ is required to achieve control. Note that, the total number of control points necessary for achieving control is dependent on the overall system size, with smaller systems requiring a sparser array of control points.

Controlling Spiral Turbulence

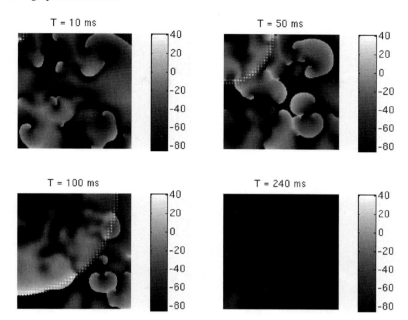

Fig. 4 Control of spatiotemporal chaos in the 2-dimensional Luo-Rudy I model of linear dimension $L = 9$ cm. The control points are arranged in a square grid with a spacing of $d = 0.225$ cm. A control stimulus of strength $A = 75\,\mu\text{Amp/cm}^2$ is applied at each control point for an average duration of 5 ms. The control was initiated at $T = 0$ ms

Since all real biological systems have depth, our control scheme must also be able to successfully eliminate all activity in a 3-dimensional system, if it is to be used as a practical defibrillation method. To this end, we consider a 3-dimensional Panfilov system and generate spatiotemporal chaos by initiating scroll-wave breakup. We then apply the control scheme through an array of electrodes which are in contact with only one face of the system, since we cannot physically penetrate the tissue with electrodes. For a system with a depth of 0.4 cm, we observe that using two waves of control stimulation, the second wave following the first after 165 ms, all spatiotemporally chaotic activity is eliminated. However, for systems with larger depth, the control scheme was not as effective. The success of our method for simulated tissue with lower values of thickness suggests that it could be very effective in terminating atrial fibrillation.

All the previous examples where the control was carried out were for homogeneous media. Actual cardiac tissue often contains many inhomogeneities, such as, non-excitable scars or blood vessels, as well as regions with significantly different restitution properties. Therefore, we next probe the effectiveness of our control scheme in the presence of large tissue heterogeneities. For this, we insert a square region with linear dimension $L_{\text{inhomogeneity}} = 5.5$ cm having a much lower diffusion constant (and hence, conduction speed) in the 2-dimensional Panfilov model with linear dimension $L = 12.8$ cm. As before, once the spatiotemporal chaotic state is

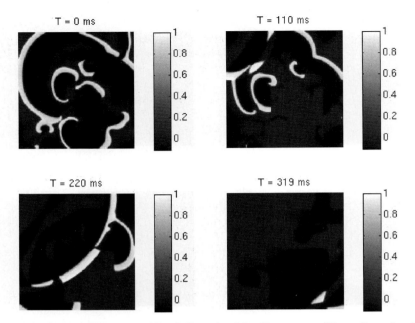

Fig. 5 Spatiotemporal chaos control in a 2-dimensional Panfilov system of linear dimension $L = 12.8$ cm with an inhomogeneity of linear dimension $L_{\text{inhomogeneity}} = 5.5$ cm. The control array spacing $d \to 0$ and the strength of control stimulus is $A = 1.5$ dimensionless units applied at each point for an average duration of 15 ms

established, the control wave is started from a corner of the system. We observe that the control scheme is successful in eliminating all activity within 350 ms (Fig. 5).

4 Discussion

The basic idea behind the proposed control scheme is that waves in excitable media mutually annihilate when they collide with each other. Therefore, if we activate the regions immediately in front of all excitations in a spatiotemporally chaotic system, the system should quickly return to the quiescent state. In principle, this is possible by using a wave initiated at a single point. However, after engaging the nearest excitation, such a control wave would be broken up (if not terminated altogether) and will be unable to successfully engage the other remaining excitations in the system. In fact, this is the reason why spatiotemporal chaos requires spatially extended control for its termination (except for the special case of overdrive pacing). We, therefore, use multiple control points to reinforce the control wave at regular intervals. Although locally the method works as described above for a single wave, i.e., colliding with and annihilating chaotic excitations, it still manages to impose control over the entire domain. This is because, although the control wave is distorted every time such an interaction occurs, the wave-shape is restored by the control array.

Controlling Spiral Turbulence

The above discussion naturally leads to a consideration of the role of the propagation velocity of the control wave, $v_{control}$. Note that, under the limiting case of $v_{control} \to 0$, the situation is identical to the case of local control from a single point located at a corner of the simulation domain. As expected, the resulting wave can interact effectively only with the nearest spiral wave and the control stimulation is unable to engage all excitations resulting in failure of termination. The other extreme case of $v_{control} \to \infty$ corresponds to the case of applying the control stimulus simultaneously at all points in the array. This reduces to global control if the array spacing, $d \to 0$. In this very special case of control being applied on all cells, the efficacy of the control method is not dependent on $v_{control}$. In fact, the global scheme is found to terminate chaos faster than the propagating control wave method. However, for $d > 0$, we observe that applying simultaneous control at all control points often fail (Fig. 6), whereas, using a finite propagation velocity is successful in eliminating all activity. This suggests the existence of an optimal range for the propagation velocity $v_{control}$.

In addition to the control wave velocity, the other parameter which is a critical deciding factor for successful control is the amplitude of the control stimulus, A. For example, in the 2-dimensional Panfilov model ($L = 12.8$ cm), the control method fails even with $d \to 0$ if the control stimulus amplitude is less than the critical value

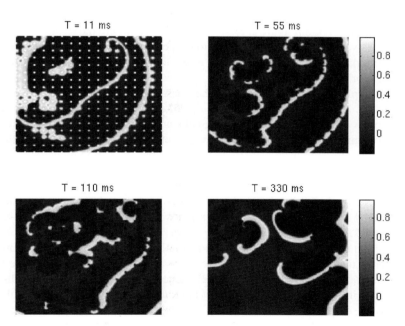

Fig. 6 Simultaneous stimulation of control points with an array spacing of $d = 0.8$ cm fails to eliminate the spatiotemporally chaotic activity in a 2-dimensional Panfilov model having linear dimension $L = 12.8$ cm. The control stimulus has amplitude $A = 10$ (dimensionless units) and applied for a duration of 17.93 ms at every point. Note that, using the same control power, our proposed method is successful in terminating all activity

$A_c = 1.3$. To understand the reason why such a critical value exists we consider in detail the process by which the control method terminates excitation. On application of control at a particular point, the corresponding cell is excited if it is in the resting state. However, if the cell is in the refractory state, whether it will be excited or not depends on the control amplitude. Large A would allow cells in the relative refractory state to be activated. The degree of refractoriness of such cells is quantified in the Panfilov model by the recovery variable g, and the magnitude of A decides whether a cell with a specific value of g can be excited. When $A < A_c$, the control stimulus is unable to affect cells which are refractory but have low g values, so that they recover very soon after the passage of the control wave. As a result, surviving fragments of chaotic excitation may persist for just long enough to be able to excite such neighboring recovering regions through diffusion. This will result in failure of termination. To prevent this, we need to ensure that the recovery time for a refractory cell exceeds the duration of survival for the chaotic excitations on being subjected to the control stimulation. With a sufficiently high A, all regions which might have recovered during the time up to which chaotic activity survives, are excited. As a result, there are no regions into which the chaotic activity can diffuse.

Figure 7 shows the result of applying control with stimulus amplitudes above and below the critical value. The top figure is for the case of successful termination when $A > A_c$ and we note that a cell is either excited by the control stimulation, or, if unaffected, then is quiescent even in the presence of diffusion from neighboring excited cells. The bottom figure corresponds to $A < A_c$ where the termination fails, and we observe that, under certain conditions, a cell may remain unaffected by the control stimulation but then gets excited a few tens of ms later through diffusion from neighboring chaotic regions.

This is explicitly shown in terms of the phase space dynamics of the Panfilov model in Fig. 8. For the case when $A > A_c$, all cells which could have been activated through diffusion from neighboring cells are forced into the excited state by the control stimulation. Cells, whose degree of refractoriness are such that they are unaffected by diffusion have $e < e^*$, and do not show any action potential even on application of the control stimulus. On the other hand, when $A < A_c$, while cells which have $e > e^*$ are activated by the control, we find that cells with $e < e^*$ at the time instant when control stimulation is applied do not show any response but are excited by diffusion from neighboring regions sometime later when their e-value rises above e^*.

In addition to the propagation speed and stimulus amplitude, the other important parameter of the proposed control scheme is the array spacing, d. We observe that as d is gradually increased, a situation is reached when the control fails. This is expected, as in the extreme limit when d approaches the linear dimension of the system, L, the control method reduces to a local stimulation scheme which is known to fail for the low frequency of stimulation that we use. Also, as noted before, when d is reduced to 0, the scheme is identical to the global control method which requires high power. Therefore, we can infer the existence of an optimal range of values for d at which control is successful but without the necessity of using a large number of control points. In Fig. 9 we show, for the realistic Luo-Rudy I model, the dependence

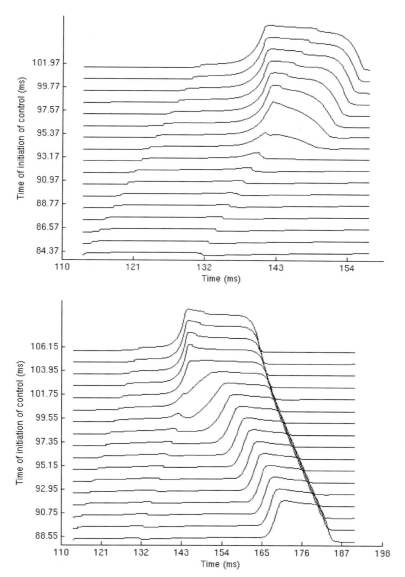

Fig. 7 The time-evolution of trans-membrane potential for a cell located at ($x = 0.45$ cm, $y = 1.35$ cm) from the *top left corner* of a 2-dimensional Panfilov model having linear dimension $L = 12.8$ cm, for different times of initiation of the control wave (i.e., the activation of the electrode located in the *top left corner*). The top figure corresponds to control amplitude $A = 1.5 > A_c$ and the *bottom* figure to $A = 1.0 < A_c$ where A_c ($= 1.3$) is the critical value of the control signal amplitude. The control stimulation is applied at each point for an average duration of 15 ms.

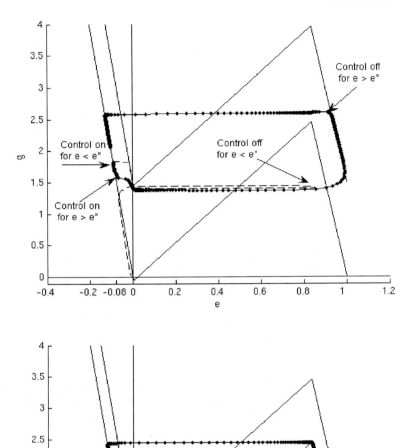

Fig. 8 Phase space for a single cell in the 2-dimensional Panfilov model showing the effect of control stimulation with amplitude (*top*) $A = 1.5$ and (*bottom*) $A = 1.0$, applied at each point for an average duration of 15 ms. The broken line is for the case when the value of the fast variable of the cell at the time of stimulation, $e < e^*$ and the dotted line is for $e > e^*$. Here, $e^* = -0.08$ is the critical value of e below which diffusion from neighboring excited cells fail to excite the cell under consideration

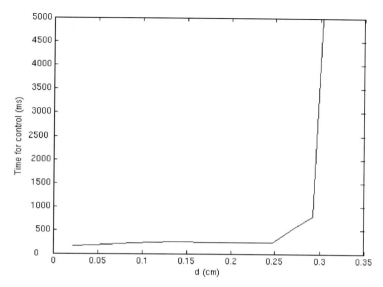

Fig. 9 Time required to control spatiotemporal chaos in a 2-dimensional Luo-Rudy I system having linear dimension $L = 9$ cm, shown as a function of the control array spacing, d. For $d < 0.25$ cm, chaotic activity is eliminated within 250 ms whereas for $d > 0.3$ cm, termination has not been achieved up to 10 s, the longest duration for which we have carried out the simulation

of the time required to terminate spatiotemporal chaos on the array spacing d. We observe that the time diverges around $d = 0.3$ cm, indicating the failure of the control method. Similar behavior is observed for the Panfilov model.

5 Conclusions

As indicated in our review of proposed control techniques, there have been many attempts at using nonlinear dynamics tools to develop a low-amplitude stimulus scheme for terminating spatiotemporal chaos in excitable media which characterizes VF. However, for practical implementation, all such methods would have to be capable of being applied through an Implantable Cardioverter Defibrillator (ICD). Such devices are currently the standard preventive treatment against VF, being regularly implanted in many patients through a relatively safe surgical procedure. It is used to monitor the heart rhythm of a patient diagnosed to be at high risk of fibrillation and applies electrical shocks, when necessary, through one or few electrodes placed on the heart wall. Any control scheme developed for terminating spatiotemporal chaos in the heart would have to conform to the limitations of the ICD, the principal one being the use of a small number of electrodes. This obviously overrules the global control schemes from the domain of practical applicability. While local control schemes might seem to satisfy the above constraints, the use of extremely high stimulation frequencies in overdrive pacing means that such methods are unstable and therefore potentially dangerous, as discussed earlier.

We believe that the method proposed in this paper avoids these disadvantages of the global and local control methods. By using a traveling wave of control stimulus that uses low-amplitude stimulation at each control point, we do away with the need for large defibrillation currents that are conventionally used. Further, our scheme does not require overdrive pacing and is therefore, robust even in the presence of large tissue inhomogeneities. Moreover, by using spatially sparse control points, the proposed method ensures that the power required for terminating chaos is less than that used by global low-amplitude control methods. As biological systems like the heart have depth, any realistic control scheme must necessarily be successful in the bulk of a three dimensional system while being applied only at the surface, a requirement which our scheme fulfills. We have also verified that the method is effective in controlling simulated arrhythmic episodes in realistic models of cardiac tissue incorporating several types of ion channels. These points suggest that the proposed control scheme marks a step forward in devising a low-amplitude control scheme that can be implemented in a conventional ICD.

Acknowledgments We thank the IFCPAR (Project No. 3404-4) and the IMSc Complex Systems project for financial support. Computations were carried out in the NCSC and Vindhya machines at IMSc. We are grateful to A. Pumir and J. Zebrowski for helpful suggestions.

References

1. R. N. Anderson, K. D. Kochanek, S. L. Murphy, *Report of Final Mortality Statistics 1995*, NCHS Monthly Vital Statistics Report **45**, Supplement 2, 7 (1997)
2. J. Keener, J. Sneyd, *Mathematical Physiology*, Springer, Berlin (1998)
3. A. V. Panfilov, P. Hogeweg, *Spiral breakup in modified FitzHugh-Nagumo model*, Phys. Lett. A **176**, 295–299 (1993)
4. A. V. Panfilov, *Spiral breakup as a model of ventricular fibrillation*, Chaos **8**(1), 57–64 (1998)
5. A. Hodgkin, A. Huxley, *A quantitative description of ion currents and its application to conduction and excitation in nerve membranes*, J. Physiol. (London) **117**, 500–544 (1952)
6. C. H. Luo, Y. Rudy, *A model of the ventricular cardiac action potential*, Circ. Res. **68**(6), 1501–1526 (1991)
7. R. Pandit, A. Pande, S. Sinha, A. Sen, *Spiral turbulence and spatiotemporal chaos: Characterization and control in two excitable media*, Physica A **306**, 211–219 (2002)
8. F. X. Witkowski, L. J. Leon, P. A. Penkoske, W. R. Giles, M. L. Spano, W. L. Ditto, A. T. Winfree, *Spatiotemporal evolution of ventricular fibrillation*, Nature (London) **392**, 78–82 (1998)
9. R. A. Gray, A. M. Pertsov, J. Jalife, *Spatial and temporal organization during cardiac fibrillation*, Nature (London) **392**, 75–78 (1998)
10. P. Wang, P. Xie, H. Yin, *Control of spiral waves and turbulent states in a cardiac model by travelling wave perturbations*, Chin. Phys **12**, 674–682 (2003)
11. G. V. Osipov, J. J. Collins, *Using weak impulses to suppress traveling waves in excitable media*, Phys. Rev. E **60**, 54–57 (1999)
12. R. A. Gray, *Termination of spiral wave breakup in a Fitzhugh-Nagumo model via short and long duration stimuli*, Chaos **12**, 941–951 (2002)

13. S. Alonso, F. Sagues, A. S. Mikhailov, *Taming Winfree turbulence of scroll waves in excitable media*, Science, **299**, 1722–1725 (2003)
14. S. Alonso, F. Sagues, A. S. Mikhailov, *Periodic forcing of scroll rings and control of Winfree turbulence in excitable media*, Chaos, **16**, 023124 (2006)
15. S. Alonso, J. M. Sancho, F. Sagues, *Suppression of scroll wave turbulence by noise*, Phys. Rev. E **70**, 067201 (2004)
16. A. T. Stamp, G. V. Osipov, J. J. Collins, *Suppressing arrhythmias in cardiac models using overdrive pacing and calcium channel blockers*, Chaos **12**, 931–940 (2002)
17. H. Zhang, B. Hu, G. Hu, *Suppression of spiral waves and spatiotemporal chaos by generating target waves in excitable media*, Phys. Rev. E **68**, 026134 (2003)
18. J. Breuer, S. Sinha, *Controlling spatiotemporal chaos in excitable media by local biphasic stimulation*, preprint nlin.CD/0406047 (2004)
19. H. Zhang, Z. Cao, N. Wu, H. Ying, G. Hu, *Suppressing Winfree turbulence by local forcing excitable systems*, Phys. Rev. Lett. **94**, 188301 (2005)
20. A. V. Panfilov, J. P. Keener, *Effects of high frequency stimulation on cardiac tissue with an inexcitable obstacle*, J. Theor. Biol. **163**, 439–448 (1993)
21. S. Sinha, A. Pande, R. Pandit, *Defibrillation via the elimination of spiral turbulence in a model for ventricular fibrillation*, Phys. Rev. Lett. **86**, 3678–3681 (2001)
22. D. J. Gauthier, G. M. Hall, R. A. Oliver, E. G. Dixon-Tulloch, P. D. Wolf, S. Bahar, *Progress toward controlling in vivo fibrillating sheep atria using a nonlinear-dynamics-based closed-loop feedback method*, Chaos **12**, 952–961 (2002)
23. A. R. J. Mitchell, P. A. R. Spurrell, L. Cheatle, N. Sulke, *Effect of atrial anti-tachycardia pacing treatments in patients with an atrial defibrillator: Randomised study comparing sub-threshold and nominal pacing outputs*, Heart **87**, 433–437 (2002)

Suppression of Turbulent Dynamics in Models of Cardiac Tissue by Weak Local Excitations

E. Zhuchkova, B. Radnayev, S. Vysotsky and A. Loskutov

Abstract On the basis of a modified FitzHugh-Nagumo system and a quite realistic ionic Fenton-Karma model describing the wave propagation in cardiac tissue we resolve the problem of suppressing the fibrillative activity of the heart by a low-voltage local electrical forcing. Such a low-energy defibrillation has a great advantage in comparison with other widespread methods since it, in particular, does not require the knowledge of the frequency of re-entrant waves. All the rotating waves are suppressed almost simultaneously and the initial cardiac rhythm can be reestablished because after stabilization the medium goes to a spatially homogeneous steady state.

1 Introduction

The suppression of the turbulent dynamics of excitable media, which appears through a set of coexisting spiral waves, by means of a small periodical (almost) point action is a very important area of current investigations in view of application to cardiology.

Excitation waves in the cardiac tissue, so-called action potentials, originate in the sinoatrial node (SA) and spread successively over the right atrium and the left atrium. Then they pass through the atrioventricular node (AV), bundle of His and Purkinje fibers, and finally reach the walls of the right and left ventricles. The normal rhythm of the heart is determined by the activity of the SA node which is called the leading pacemaker (a source of concentric excitation waves) or the first order driver of the rhythm. In addition to the SA node cells, the other parts of the cardiac conductive system can reveal automaticity. So, the second order driver of the rhythm is located in the AV conjunction. The Purkinje fibers are the rhythm driver of the third order.

Any abnormality in the cardiac rhythm is said to be arrhythmia. It can develop by several reasons, the dominant of which consists of the change in the intrinsic

A. Loskutov (✉)
Moscow State University, Moscow 119992, Russia
e-mail: loskutov@chaos.phys.msu.ru

properties of the excitable tissue. In this case the destruction of the excitation wave fronts is possible such that the fibrillation phenomenon may appear.

The fibrillation is the prevalent mode of death among patients with cardiovascular diseases. The extreme form of cardiac fibrillation is ventricular fibrillation (VF), which is a fast developing disturbance of spatially organized contraction of ventricles. The dominating hypothesis in the current theory of excitable systems is that fatal cardiac arrhythmias, fibrillations, occur due to the creation of numerous autowave sources, so-called re-entry, which are spiral waves or vortex structures (i.e., spatiotemporal chaos, see, e.g., [1, 2] and references therein), in cardiac tissue.

To suppress the heart fibrillation, the application of high-energy electrical stimulation through the patient's chest is commonly used. However, high-energy shock can cause the necrosis of myocardium or give rise to functional damage manifested as disturbances in atrioventricular conduction. So, it is necessary to find another method of stimulation of the fibrillative heart.

The application of electrical pulses for the termination of fibrillations is also used in implantable cardioverter defibrillators (ICDs) initiating low-power electrical pacing pulses automatically when they detect dangerous activity. However, the ICD action is very painful and occurs even in the case of complex arrhythmias that are not associated with fibrillation. Therefore, a very important factor in the design of modern ICDs is reduction of the stimulation amplitude in order to avoid painful high-energy shocks and damage to the heart itself and surrounding tissues. Thus, there is a high demand in clinics on alternative methods of defibrillation, which would work with lower voltages.

The recent research [3] may provide an alternative to the conventional ICD therapy by terminating re-entrant arrhythmias with the field strengths that are 5–10 times (or delivered energy 25–100 times) weaker than usual defibrillation shocks. However, these methods (so-called unpinning) are valid only for the high-risk cardiac patients who had, in particular, previous myocardial infarctions ("heart attacks").

Theoretical studies suggest that low-energy defibrillation protocols are also possible at exploiting the dynamical properties of re-entrant waves under electrical forcing, known as feedback-driven resonant drift [4]. However, in practice the major problem is the change of the resonant frequency with the position of a rotating wave, especially close to unexcitable boundaries.

The qualitatively different method of a low-amplitude suppression of the chaotic dynamics is realized by parametric perturbations without feedback. Firstly it was proposed in [5] and mathematically substantiated in [6]. Although both parametric suppressing (non-feedback) and controlling (feedback) lead to the stabilization of complex dynamics, they are not realizable for the electrical defibrillation. The only application seems to be in formulating drug therapies which modify ionic currents in order, for example, to prevent alternans (beat-to-beat alternation in the action potential duration), which is presumably one of the causes of breakup of a single rotating wave into multiple re-entry (see, e.g. [7–9]).

The recent investigations of active media offer new opportunities for the electrical defibrillation: The amplitude of the external stimulation can be *essentially* decreased and the turbulent regime in excitable systems can be suppressed by a

Fig. 1 The scheme of a weak local stimulation of the turbulent medium

sufficiently weak non-feedback periodic external forcing applied globally [10, 11] or locally [12–15] (see also references therein). By these manners, it is possible to stabilize the turbulent dynamics and reestablish the initial cardiac rhythm, because such a strategy leads to the relaxation of the medium to the rest state.

In the present paper we show that stabilization of fibrillative dynamics of the cardiac tissue (in other words, suppression of spatio-temporal chaos in the excitable medium) can be achieved by weak local excitations (Fig. 1). Our defibrillation scheme is realized by means of a modified FitzHugh-Nagumo model and simplified ionic model (SIM) of the cardiac action potential, so-called Fenton-Karma equations. All the simulations are performed in two dimensions.

2 The FitzHugh-Nagumo-Type Model and the Fenton-Karma Equations

The FitzHugh-Nagumo-type system can be written as follows:

$$\frac{\partial u}{\partial t} = \Delta u - f(u) - v, \qquad (1)$$
$$\frac{\partial v}{\partial t} = g(u, v)(ku - v).$$

To get correlation with the heart tissue, usually the following admissions are applied:

$$f(u) = \begin{cases} C_1 u, & u < u_1, \\ -C_2 u + a, & u \in (u_1, u_2), \\ C_3 (u - 1), & u > u_2, \end{cases} \quad g(u, v) = \begin{cases} G_1, & u < u_1, \\ G_2, & u_1 > u_2, \\ G_3, & u < u_1, v < v_1. \end{cases} \qquad (2)$$

Here u and v are activator and inhibitor variables, respectively. The parameter values are $C_1 = 20$, $C_2 = 3$, $C_3 = 15$, $u_1 = 0.0026$, $u_2 = 0.837$, $v_1 = 1.8$, $k = 3$, $G_2 = 1$. Parameters $G_1 \in [1/75, 1/30]$ and $G_3 \in [0.1, 2.0]$ remain to be free.

One of the advantages of this model is the presence of two independent relaxation parameters. One of them, G_3, takes into account a relative relaxation period for small values u and v. The other one, G_1, gives an absolute relaxation period for large values of v and intermediate values of u that corresponds to the leading and trailing fronts of the wave. In spite of its simplicity, this model describes real experimental data sufficiently well even for the myocardium tissues of mammals [16]. For example, it correctly reproduces the shape of the action potential when varying the parameters and initial conditions in wide intervals and can demonstrate all types of structures inherent in an excitable tissue.

Numerical simulations were carried out in two-dimensional (2D) grids of 175 × 175 and 350 × 350 elements with periodic (i.e. on a torus) boundary conditions. This geometry is quite adequate to an intact heart in contrast to a sheet of tissue (Neumann boundary conditions) and allows to exclude wave attenuation.

The more sophisticated Fenton-Karma equations have the following form [17, 18]:

$$\partial_t u = \nabla(D\nabla u) - (J_{fi}(u, v) + J_{so}(u) + J_{si}(u, w)),$$
$$\partial_t v = \Theta(u_c - u)(1 - v)/\tau_v^-(u) - \Theta(u - u_c)v/\tau_v^+,$$
$$\partial_t w = \Theta(u_c - u)(1 - w)/\tau_w^- - \Theta(u - u_c)w/\tau_w^+, \qquad (3)$$

where u is a dimensionless membrane potential; v, w are fast and slow ionic gates, respectively; D is a diffusion tensor that in our case is a diagonal matrix with equal diagonal elements (0.001 cm^2/ms), which corresponds to an isotropic medium; J_{fi}, J_{so}, J_{si} are scaled ionic currents describing the Na$^+$, K$^+$, Ca^{2+} currents, respectively; $\Theta(x)$ is a standard Heaviside step function (other functions, equations for the ionic currents and parameter values see in [18]). The parameters correspond to the steep action potential duration (APD) restitution, i.e. steep dependence of APD on diastolic interval (DI), or time between successive excitations (fitted to accurately represent the APD restitution in the full Beeler-Reuter model, so-called Set 3 in [18]) with breakup close to the tip.

The SIM supports many different mechanisms of the spiral wave breakup into a complex re-entrant activity. Our numerical simulations were performed in a 2D grid of 500 × 500 elements corresponding to the tissue size of 12.5 × 12.5 cm. In this case we also used periodic boundary conditions, i.e. the torus topology.

Analyzing the spatially extended systems, the main problem is to obtain correctly the characteristics of spatio-temporal chaos. As it was noted above, for the excitable medium the spatio-temporal chaos appears via a set of rotating waves. As is known, each of them has a phase singularity (PS), i.e. tip, or filament [19]. In the present paper, to estimate the depth of the irregularity in the medium dynamics we use the number N of phase singularities.

The method proposed in [19] is based on the fact that the tip of the spiral wave (as well as any point of discontinuity of the wave front) is a singularity for the phase field $\varphi(x, y, t) = \arctan 2(U(x, y, t) - U^*, V(x, y, t) - V^*)$. In this case the quantity

$$n = \frac{1}{\pi} \int \nabla\varphi \, dl$$

which is called the topological charge, is not equal to zero only if such a singularity is located within the integration contour. In this case, n is an integer, whose sign determines the chirality of the spiral wave.

This approach allows us to attain more effective control over the system dynamics. Its main advantage consists of the availability of the well-developed algorithms. Our investigations showed that the number of phase singularities may serve as a rather simple and the visual estimation of the efficiency of the turbulence suppression.

In Fig. 2 the number of the phase singularities of spiral waves as a function of time in the FitzHugh-Nagumo system is shown. The turbulent regime appears from a destroyed single spiral wave. This figure corresponds to the spatial pattern shown in Fig. 3. During our investigations we found that the initial distribution taken in the form of the plane half-wave is transformed into spatio-temporal chaos after about 2000 time units. Such a regime was used for further analysis as the initial state of the number system when studying the possibility of the suppression of turbulent dynamics.

For the Fenton-Karma equations we have found that an initial spiral wave breaks into complex turbulent pattern after approximately 450/470 ms. The PS as a function of time is shown in Fig. 4. Development of turbulent dynamics and creation of a set of coexisting spiral waves is shown in Fig. 5.

The described turbulent states of the SIM were considered as initial ones for all our suppression attempts. In turn, to suppress the spatio-temporal turbulence it is necessary to determine shapes, amplitudes, durations and frequencies of stimulations.

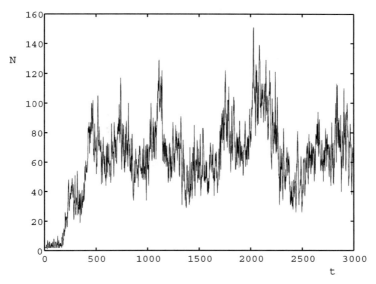

Fig. 2 The number of phase singularities of spiral waves in the FitzHugh-Nagumo system (1) and creation of the turbulent regime: $G_1 = 0.01$ and $G_3 = 0.5$

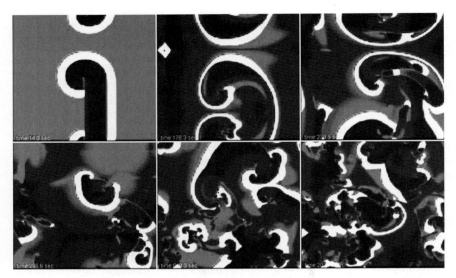

Fig. 3 Destruction of a spiral wave and creation of spatio-temporal chaos in the FitzHugh-Nagumo system (1)–(2)

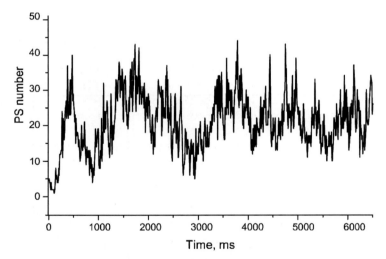

Fig. 4 The number of PSs as a function of time for the SIM (3)

3 Suppression of the Fibrillative Activity

First, let us consider the important problem of the detection of the perturbation frequencies that provide the effective stabilization. Because the search of the suppressing frequencies at random is ineffective, we used the method that allowed us to find the frequency intervals ensured the high efficiency of the chaos suppression.

Suppression of Turbulent Dynamics

Fig. 5 Development of the fibrillative regime in the SIM (3) from an initial single spiral wave

The idea of our method consists of the known properties of active media (see [20]). For the observation of the chaos suppression phenomenon it is necessary that the frequency of circular pacemaker waves in the medium is close to the maximal possible frequency for the given system parameters. To solve this problem, one can measure the period of target waves emitted by an external pacemaker (electrode) as a function of its own period and then choose values in the frequency intervals near the maxima of the frequency dependence.

As is known, if there are several co-existing sources of periodic waves in an excitable medium, the interaction of waves leads to the suppression of the sources with a longer period by a source with a shorter period [4]. This is caused by the destructive interaction of colliding waves in active media, which mutually annihilate. If the leading source is an external electrode, it can suppress re-entry subject to the correct choice of its frequency and shape of stimulation. Note that elimination of re-entrant waves is not only connected with the drift of spiral waves, because periodic boundary conditions are used.

3.1 The FitzHugh-Nagumo System

To find effective frequencies for the suppression of the turbulence in the FitzHugh-Nagumo model (1)–(2), we generated pacemakers in the two-dimensional medium volume and determined the frequency ω_{cw} of the obtained waves as a function of the pacemaker frequency ω_{ef} (Fig. 6). One can expect that in the frequency intervals

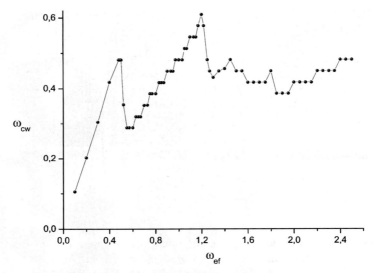

Fig. 6 The frequency ω_{cw} of circular waves in the medium as a function of the external biphasic force frequency ω_{ef}, $G_1 = 1/30$, $G_3 = 1.0$

near the maxima of this dependence the spatio-temporal chaos can be suppressed by mild point external perturbation.

The shape of external stimulation is also one of the key factors strongly influencing on the suppression effectiveness. The defibrillation shocks used in clinical practice are of rectangular monophasic and biphasic shapes.

To stabilize the spatio-temporal chaotic dynamics, in contrast to the defibrillation by single pulses, applied to the entire muscle or a quite large part of it, we added the external *periodic* forcing $I(t)$ of the same bi- and monophasic waveforms (Fig. 7) with the frequency ω_{ef} and the amplitude A to a *point* 2×2 nodes of a medium, thus symbolizing an external electrode. In other words, to the first equation of the system (1)–(2) we added periodic impulses with period T of two different types: $I_{-+}(t) = A(2\theta(t-T\tau)-1)$ and $I_{+}(t) = A\theta(t-T\tau)$, where θ is the Heaviside step function and τ is varied between 0.1 and 0.9.

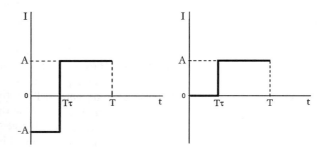

Fig. 7 The shapes of the excitation impulses: $I_{-+}(t)$ (*left*) and $I_{+}(t)$ (*right*)

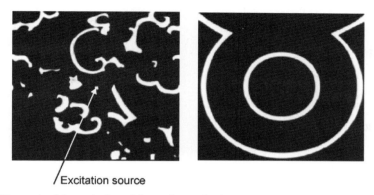

Fig. 8 The result of the excitation of the medium point 2 × 2

The result of the point action on the turbulent active medium is shown in Fig. 8. The number N of phase singularities as a function of time in the system with the external point perturbation ($G_1 = 1/30$, $G_3 = 1.0$, $A = 6$) is shown in Fig. 9a,b. In these figures the starting points of the curves correspond to the established turbulent regime of the medium.

One can see that for $\omega_{ef} = 1.2$, which exactly corresponds to the maximum of the function shown in Fig. 6, the suppression efficiency is very high, and during a quite short period of time the external pacemaker completely eliminates all the spiral waves. However, at $\omega_{ef} = 0.48$ that corresponds to the lower maximum it is not possible to suppress the turbulence during the observation time.

As to the impulse shape, it is necessary to say the following. We found that the suppression phenomenon exists only in the sufficiently narrow interval $\tau \in (0.7, 0.75)$. In other words, the impulses should be short enough. Simultaneously, for the excitation form I_+ the suppression time is approximately twice as much as for I_{-+}.

3.2 The Fenton-Karma Model

For the SIM (3) the same mono- and biphasic waveforms with the frequency ω_{in} and the amplitude A applied to a *point* (2 × 2 nodes) of a medium were chosen. Now it is necessary to find correctly amplitudes and durations of the stimulation. The duration of impulses depends on both period T and parameter $\tau (0 < \tau < 1)$ as $T\tau$ and $T(1-\tau)$.

Since the suppression of spiral waves is possible only at high frequencies (we consider periods T less than 500 ms), thus duration is determined by τ. To find the highest possible values of A and τ corresponding to experimentally observed shock-induced variations of the membrane potential, we applied single stimuli of various amplitudes and durations. We found that optimal values are the following: $\tau = 0.05/0.15$ for monophasic and $\tau = 0.25/0.3$ for biphasic stimuli. As an example we took $\tau = 0.1$ for the monophasic stimulation and $\tau = 0.3$ for the

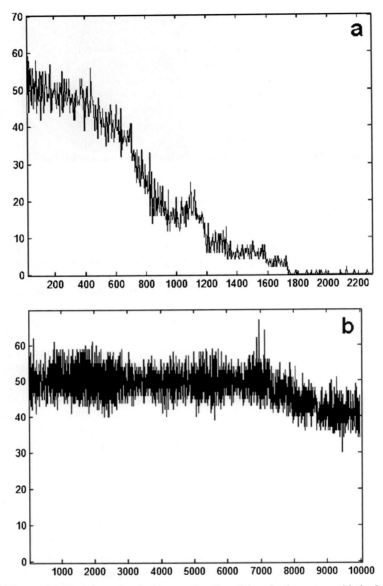

Fig. 9 The number N of phase singularities as a function of time for the system with the following parameters: $G_1 = 1/30$, $G_3 = 1.0$, $A = 6$, $\omega_{ef} = 1.2$ (**a**) and $\omega_{ef} = 0.48$ (**b**)

biphasic waveform. For the stimulation amplitude we chose $A = 10\,\mu\text{A}/\text{cm}^2$ for both waveforms. This value corresponds to the fourfold excitation threshold. Larger amplitudes of the stimulation and durations produce shock-induced variations of the membrane potential much larger than the maximal observed value [21].

We have measured the APD restitution curves obtained by two successive $S1$ and $S2$ (so-called $S1S2$ protocol, see, e.g. [17]) stimuli of various amplitudes and durations (we used the 80% cutoff when calculating restitutions). It was found that the APD restitution curves are the same for all stimuli and exactly look like the curve shown in Fig. 4 in [18]. However, to get the smallest DI of 43 ms, the $S1$-$S2$ interval of various stimuli was different. It is less when the stimulation amplitudes and/or durations are greater.

As was described in the previous Section, to select frequencies of both stimulations we generated pacemakers in a quite small volume of the medium and determined the frequency ω_{out} of the target waves as a function of the internal pacemaker frequency ω_{in} (Fig. 10).

First, we tried to suppress complex activity by the monophasic stimulation with $\omega_{\text{in}} = 3.13\,\text{Hz}$ and $7.0\,\text{Hz}$, corresponding to the first two frequency maxima on the left-hand side of Fig. 10. Note that stimulating at high frequencies (when $T_{\text{in}} < T_{\text{R}}$, where T_{in} and T_{R} are the stimulation period and the refractory period of the medium, respectively) leads to Wenckebach rhythms which in turn may cause the re-entry. Hence, in practice it makes a sense to apply stimulations with a maximum possible value of the frequency $\omega_{\text{in}} = \omega_{\text{out}}$ that provides $T_{\text{in}} > T_{\text{R}}$ (the first maximum for both waveforms in Fig. 10).

We found that although the suppression phenomenon was observed for both pacemaker frequencies, it depends on the suppression onset. For example, the monophasic forcing of $\omega_{\text{in}} = 7\,\text{Hz}$ leads to the stabilization of chaotic dynamics if it starts at 500 ms, but suppression is unsuccessful if the suppression onset is 600 ms (see Fig. 11, solid and dashes curved, respectively). Vice versa, the stimulation with $\omega_{\text{in}} = 3.13\,\text{Hz}$ started at 600 ms was successful. This phenomenon is connected with the nonlinearity of the system and the phase of the external excitation.

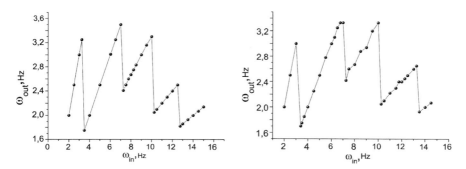

Fig. 10 The frequency ω_{out} of target waves as a function of the pacemaker frequency ω_{in}. Left-hand side corresponds to the monophasic stimulation, right-hand side corresponds to the biphasic stimulation

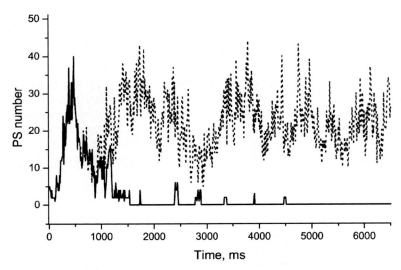

Fig. 11 The number of PSs as a function of time for the SIM during the monophasic stimulation with $\omega_{in} = 7.0$ Hz and $A = 10\,\mu A/cm^2$. Suppression onsets are: t = 500 ms (*solid curve*), 600 ms (*dashed curve*)

The further careful analysis showed that the dependence on the suppression onset (and on the place of the excitation) can be eliminated if we use slowly moving pacemaker(s) (see below Section 4).

Second, we forced the system by the biphasic stimulation with $\omega_{in} = 3.13$ and 7.25 Hz corresponding to the first two frequency maxima on the right-hand side of Fig. 10. But, in contrast to the monophasic stimulation, biphasic forcing leads to the stabilization of turbulent dynamics by stimuli with $\omega_{in} = 3.13$ Hz started at 500 ms and by the stimulation with $\omega_{in} = 7.25$ Hz started at 600 ms. Because the frequency interval corresponding to the second maximum on the right-hand side of Fig. 10 is quite wide, it is rather complicated to select the appropriate value of the stimulation frequency. Figure 12 shows the susceptibility to its choice. It represents the time evolution of the excitation pattern during the biphasic stimulation started at 600 ms with close frequencies 7.25 Hz (panel A), 7.0 Hz (panel B) and 6.75 Hz (panel C). Panel A corresponds to the successful suppression (just target waves of an external pacemaker exist at $t \geq 1200$ ms). In panel B the recovered turbulence is shown. Panel C corresponds to the unsuccessful suppression. As it was predicted, the patterns of the suppressed and recovered turbulence before the elimination of the spiral-wave activity were similar due to the small difference in the stimulus length, but the generation of new spirals resulting from an external pacemaker in the latter case (panel B) was unexpected.

Note that the biphasic stimulation with $\omega_{in} = 7.0$ Hz started at 600 ms (resulting in the recovered turbulence when $A = 10\,\mu A/cm^2$, see panel B) leads to the effective suppression at the doubling of the stimulation amplitude. However, attempt of the trebling of the amplitude fails (see Fig. 13). So, there is a nonlinear dependence on the excitation amplitude.

Suppression of Turbulent Dynamics

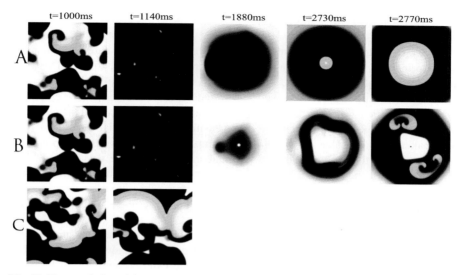

Fig. 12 Time evolution of the excitation patterns during the biphasic stimulation started at 600 ms with close frequencies. (**A**) Effective suppression of complex activity by the biphasic stimuli, $\omega_{in} = 7.25$ Hz. At the subsequent forcing all sites recovered without further activation. (**B**) Recovery of spiral-wave activity after its suppression by biphasic stimuli, $\omega_{in} = 7.0$ Hz. (**C**) Unsuccessful attempt to suppress turbulence by the biphasic stimulation, $\omega_{in} = 6.75$ Hz

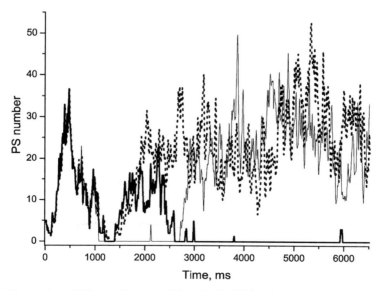

Fig. 13 The number of PSs as a function of time for the SIM during the biphasic stimulation of $\omega_{in} = 7$ Hz and $A = 10 \, \mu A/cm^2$ (*thin curve*), $A = 20 \, \mu A/cm^2$ (*thick curve*), $A = 30 \, \mu A/cm^2$ (*dashed curve*). The suppression onset is $t = 600$ ms

4 Moving External Pacemakers

Thus, we come to the conclusion that spiral wave activity in the cardiac tissue can be, in principle, suppressed by a weak local periodic excitation source. This is a very important qualitative result, which gives us the hope to resolve the defibrillation problem in the future. However, unfortunately, there are some very essential limitations, which we have mentioned before but have not discussed so far. One of such limitations is the extreme system sensitivity to the initial conditions and pacemaker (or pacemakers) location. For example, if we observe the suppression phenomenon in the system with one pacemaker and then move it, say, 10 nodes from its position and integrate the system again with the same initial conditions, the suppression phenomenon disappears and/or vice versa. This effect is related to the complexity of spatial patterns emerging in the medium as a result of the interaction of a large number of spirals.

To put it simple, let us imagine such a situation, when a pacemaker is surrounded by several vortices, in such a way, that the space around it is always occupied by wave fronts of surrounding spirals. In that case the external source will be suppressed by the completely chaotic medium, and the system dynamics will develop as if it were no external impact applied to it.

To solve this problem we propose two independent directions of further investigations: to use several pacemakers and moving pacemakers. The first and the simplest direction is increasing the number of external pacemakers in the system. Unfortunately, after some experiments we have come to the conclusion that stationary pacemakers (even a comparably large number) can not guarantee chaos suppression in our systems.

Therefore, we generalized our approach and considered the case of several moving pacemakers (2 or 4). The law of their movement used is a sinus projection $\xi = \xi_0 \sin \omega(\omega_{rot} t)$, where ξ is parallel to either vertical or horizontal axis. But even here we faced very complex system behavior. Chaos suppression efficacy depends extremely on ω_{rot} as well as pacemakers number, their initial position and the distance between them.

Let us consider, for example, a FitzHugh-Nagumo system with two pacemakers located on a *vertical line* with the distance between them 176 points. Let these pacemakers move along a *horizontal line* with the amplitude $\xi_0 = 10$ and frequency $\omega_{rot} = 2.5 \cdot 10^{-4}$. Then, for the medium with $G_1 = 0.01$, $G_3 = 0.5$, we get the suppression phenomenon during 3300 time units (see Fig. 14).

If, however, we take the same condition for *horizontally located* pacemakers which are *vertically moving*, the suppression of turbulent dynamics is not observed. It can be achieved by increasing the frequency up to, for example, $\omega_{rot} = 3.0 \cdot 10^{-4}$. However, in this case the chaotic dynamics is extruded after 9500 time units (Fig. 15).

The dependence on the position of pacemakers and the frequency of their motion remains to be carefully explored. Preliminary investigations show that this dependency is essentially nonlinear.

Suppression of Turbulent Dynamics

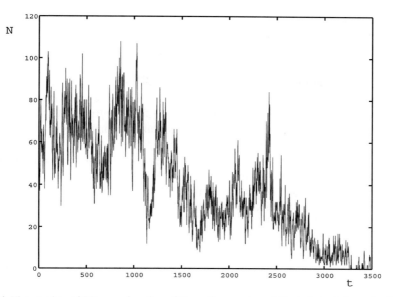

Fig. 14 The number of PSs as a function of time: Suppression of the turbulent dynamics in the FitzHugh-Nagumo type model by *horizontally moving* pacemakers

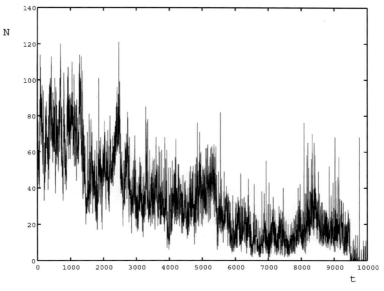

Fig. 15 The number of PSs as a function of time: Suppression of the turbulent dynamics in the FitzHugh-Nagumo type model by *vertically moving* pacemakers

5 Conclusion

The properties of spiral waves in the excitable media can be exploited for the development of novel approaches to the defibrillation of the cardiac muscle. The elimination of re-entrant arrhythmias is a very important clinical problem. That is the reason why, in the recent years a number of studies have been concentrated on the theoretical understanding of how the application of an electric current defibrillates the heart. One of our aims in present investigation is to develop new methods of internal stimulation, which reduce the defibrillation threshold.

On the basis of the FitzHugh-Nagumo and Fenton-Karma models we showed that it is quite possible to stabilize spatio-temporal chaotic dynamics by external point excitations. This approach permits to eliminate spiral waves and restore the regular behavior in the system. The main problem here is to find the excitation frequencies. But this is easily solved by the preliminary localization (Section 3).

Thus, the obtained results make possible to predict the dynamics of active systems, depending on their parameter values. Moreover, using the proposed approach one can develop a quite general theory of the chaos suppression in distributed media by point excitations.

Finally, we would like to say some words about applications to cardiology. The theory of non-linear dynamical systems could be the key for more enhanced understanding of fibrillation and its therapy. The standard method to stop chaos in cardiac muscle – defibrillation – does not distinguish normal waves and re-entry, anatomical and functional re-entry. Estimation has shown that the energy required by the new approaches can be less then the standard defibrillation energy by two orders of magnitude. This gives a good expectation for success, even if significant amount of estimated energy gain is consumed by technical limitations. A new word in the solution of the problem of sudden death due to the fibrillation phenomena is so-called implantable cardioverter defibrillators (ICDs). They are placed in bodies and realize the monitoring of cardiac rhythm. Since very important factor in the construction of the modern ICD is decreasing its mass and size, the main task is to find such an impulse form that can allow to defibrillate myocardium. Some specific waveforms of high voltage can realize low-energy defibrillation but, at the same time can induce the myocardium depression. Therewith, the energy decrease is not a unique task. In addition, it is necessary to reduce a painful feeling in the heart. This problem can be resolved by employing the difference in the nervous tissue and myocardium excitability.

Our conjecture consists of the following: To find domains of turbulence in the space of model parameters corresponding to the spiral wave solution (i.e. fully developed spatially-temporal chaos). Being in one of this domains, one has to perturb weakly a small part of the medium in a predefined way. Under such additional conditions this procedure may result in squeezing of spiral waves out of the heart to its boundaries where these waves will die out completely because they can exist only as waves in a 3D medium. In the short-time term this means, actually, defibrillation. Prolonged influence of stimulation of this kind prevents recovery of fibrillation and provides conditions for regeneration of the damaged cardiac tissue in the long-time term.

References

1. Winfree A.T. When Time Breaks Down: The Three e-Dimensional Dynamics of Electrochemical Waves and Cardiac Arrhythmias. Princeton Univ. Press, Princeton, USA, 1987.
2. Zipes D.P., Jalife J. Cardiac Electrophysiology – From Cell to Bed-Side, 2nd ed. Saunders, Philadelphia, 1995.
3. Takagi S., Pumir A., Pazo D., Efimov I., Nikolski V., Krinsky V. Phys. Rev. Lett., 93: 058101, 2004.
4. Biktashev V.N. Computational Biology of the Heart. Eds. A.V. Panfilov and A.V. Holden. Wiley, Chichester, 1997, p. 137.
5. Alekseev V.V., Loskutov A. Sov. Phys.-Dokl., 32: 270, 1987.
6. Loskutov A., Shishmarev A.I. Chaos, 4: 351, 1994.
7. Echebarria B., Karma A. Chaos, 12: 923, 2002.
8. Stamp A.T., Osipov G.V., Collins J.J. Chaos, 12: 931, 2002.
9. Allexandre D., Otani N.F. Phys. Rev. E, 70: 061903, 2004.
10. Gray R.A. Chaos, 12: 941, 2002.
11. Sinha S., Pande A., Pandit R. Phys. Rev. Lett., 86: 3678, 2001.
12. Loskutov A., Cheremin R., Vysotsky S.A. Dokl.-Phys., 50: 490, 2005.
13. Loskutov A., Vysotsky S. JETP Lett., 84: 616, 2006.
14. Zhang H., Cao Z., Wu N.-J., Ying H.-P., Hu G. Phys. Rev. Lett., 94: 188301, 2005.
15. Yuan G., Wang G., Chen S. Europhys. Lett., 72: 908, 2005.
16. Meunier C., Segev I. Handbook of Biological Physics. Elsevier, Amsterdam, 2000, vol. 4.
17. Fenton F., Karma A. Chaos, 8: 20, 1997.
18. Fenton F., Cherry E.M., Hastings H.M., Evans S.J. Chaos, 12: 852, 2002.
19. Bray M.A. et al. J. Card. Electrophysiol., 12: 716, 2001.
20. Osipov G.V., Collins J.J. Phys. Rev. E, 60: 54, 1999.
21. Fast V.G., Rohr S., Ideker R.E. Am. J. Physiol. Heart Circ. Physiol., 278: H688, 2000.

Synchronization Phenomena in Networks of Oscillatory and Excitable Luo-Rudy Cells

G. V. Osipov, O. I. Kanakov, C.-K. Chan, J. Kurths, S. K. Dana, L. S. Averyanova and V. S. Petrov

Abstract We study collective phenomena in nonhomogeneous autonomous and forced cardiac cell culture models, including one- and two-dimensional lattices of oscillatory cells and mixtures of oscillatory and excitable cells. Individual cell dynamics is described by a modified Luo-Rudy model with depolarizing current. We focus on the transition from incoherent behavior to global synchronization via cluster synchronization regimes as coupling strength is increased. These regimes are characterized qualitatively by space-time plots and quantitatively by profiles of local frequencies and distributions of cluster sizes in dependence upon coupling strength. We describe spatio-temporal patterns arising during this transition, including pacemakers, spiral waves and complicated irregular activity. We investigate the influence of external force on the processes of control of cardiac activity, e.g. on the suppression of spiral wave chaos in the network.

Keywords Cardiac tissue · spiral waves · concentric waves · chaos · synchronization · chaos suppression

1 Introduction

Processes of generation and propagation of cell excitation waves in cardiac tissues are a matter of topical interest because of their importance for understanding normal and pathological types of heart activity. Under sinus rhythm, waves of electrical activity propagate throughout the heart, eliciting a simultaneous contraction of the ventricles. It is known that a spiral wave arise from one of arrhythmias in cardiac muscle- tachycardia. Its rotation frequency is higher than a frequency of normal sequence of dirge pulse. That spiral waves break up into spiral waves chaos. And ventricular fibrillation might arise. The dynamics of heart tissues has been studied quite extensively in recent years, both experimentally and by means of numerical modeling. A special class of studies is concerned with

G.V. Osipov (✉)
Department of Radiophysics, Nizhny Novgorod University, 603950
Nizhny Novgorod, Russia

cardiac cell cultures – thin layers of cells grown in Petri dishes. Characteristic features of such systems are spontaneous oscillatory activity, spatial inhomogeneity and variability of intercellular coupling strength due to an increasing number of cell junctions. Modeling biological systems such as neuronal ensembles, kidney and cardiac tissues is one of the most rapidly developing fields of application of nonlinear dynamics nowadays. The efficiency of these methods is conditioned by the complex, though deterministic behavior of individual cells constituting the tissue. In particular, cardiac cells exhibit properties of either excitable or oscillatory systems. The former case is observed in working myocardium, and the latter is found in natural cardiac pacemakers (sinoatrial and atrioventricular nodes, Purkinje fibers). Normal heart activity is controlled by waves of excitation generated in the sinoatrial node and propagating through the conducting system and working myocardium. Deviations from the normal regime (arrhythmias) are often associated with pathological types of wave dynamics in the cardiac tissue. They include spiral waves (associated with tachycardia) and spiral wave chaos (the latter manifests itself in heart fibrillation). Significant scientific efforts have been taken to understand these regimes and develop a way of controlling them [1–7]. In the first part of the present paper we report a series of numerical experiments with one- and two-dimensional cardiac cell culture models, including inhomogeneous ensembles of oscillatory cells and mixtures of oscillatory and excitable cells. Individual cell dynamics is described by a modified Luo-Rudy model with depolarizing current. We focus mainly on the transition from incoherent behavior of uncoupled cells to global synchronization in ensembles of strongly coupled cells, when the coupling coefficient is increased from zero. This corresponds to the increase of the number of gap-junctions in the culture. We show, that this transition occurs via cluster synchronization regimes. We describe spatiotemporal patterns arising during this transition, including pacemakers, spiral waves and complicated irregular activity. These dynamical effects emerge due to spatial discreteness and inhomogeneity of the model. Similar experiments in-vitro were reported in [8]. According to [8], after approximately 24 hours of culture time, irregular spontaneous activity arises in the culture, further it organizes itself into several pacemakers emitting target waves. These pacemakers are subsequently destroyed, and spiral wave activity sets in; the number of spiral cores is changing with time [8]. In the second part, we explore the feasibility of using overdrive pacing to eliminate spiral waves and spiral wave chaos in cardiac tissue. We base our in numero experiments on theoretical principles underlying the physics of interacting waves in excitable media, and we explore a parameter space that is considerably larger than that utilized in the aforementioned experimental studies. Recently the successful use of the combination of overdrive pacing and calcium channel blockers was studied in [2]. We examine the possibility of coupling overdrive pacing with application of low-amplitude constant current. We show that weak constant current leads to transformation of spiral wave chaos into quasi-periodic, meandering spiral wave activity. We explore, in a series of computational experiments, the possibility of exploiting this effect to enhance the effectiveness of overdrive pacing in eliminating arrhythmias.

2 The Model

2.1 Excitable Cells

As a basis, we use the Luo-Rudy phase I model [9] to define the dynamics of a single cell. This model describes the dynamics of excitable cardiac cells and is defined by a system of 8 ordinary differential equations (ODE). The first of them is the charge conservation equation

$$C_m \dot{v} = -(I_{Na} + I_{si} + I_K + I_{K_1} + I_{K_p} + I_b) + I^{stimulus} \quad (1)$$

where v is the membrane voltage measured in millivolts, $C_m = 1\,\mu F/cm^2$ is the membrane capacity. The time unit is one millisecond. The ionic transmembrane currents in the right-hand part are sodium current, slow inward current (carried by calcium ions), potassium current, inward-rectifier potassium current, plateau potassium current and background Ohmic current, measured in $\mu A/cm^2$. They are defined by the following expressions

$$\begin{aligned}
I_{Na} &= G_{Na}.m^3 h j.(v - E_{Na}) \\
I_{si} &= G_{si}.df.(v - E_{si}(v,c)) \\
I_K &= G_K.xx_i(v).(v - E_K) \\
I_{K_1} &= G_{K_1}.k_{1i}(v).(v - E_{K_1}) \\
I_{K_p} &= G_{K_p}.k_p(v).(v - E_{K_p}) \\
I_b &= G_b.(v - E_b)
\end{aligned} \quad (2)$$

Here G_q and E_q with $q \in \{Na; si; K; K_1; K_p; b\}$ denote the maximal conductance and reversal potential of the corresponding ionic current. $I^{stimulus}$ is the input stimulus. The gating variables $g_i \in \{m; h; j; d; f; x\}, i = 1, \ldots, 6$, are governed each by an ODE of the type

$$\dot{g}_i = \alpha_{g_i}(v)(1 - v) - \beta_{g_i}(v)v \quad (3)$$

The 12 nonlinear functions $\alpha_{gi}(v)$ and $\beta_{gi}(v)$ as well as $E_{si}(v;c)$, $x_i(v)$, $K_{1i}(v)$, $K_p(v)$ are fitted to experimental data [9]. The dynamics of the internal calcium ion concentration c is described by an ODE of the first order

$$\dot{c} = 10^4 I_{si}(v, d, f, c) + 0.07(10^{-4} - c) \quad (4)$$

The eight ODEs (1), (3), (4) form a closed system for the variables of state v; m; h; j; d; f; x; c. The values of the constant parameters are the same as used in [2]. This model lacks many details taken into account in other models, which are much more complicated [10–13]. However, it still demonstrates good qualitative and quantitative agreement with available experimental data on single-cell dynamics [9], as opposed to other paradigmatic but more qualitative models like the FitzHugh-Nagumo model.

2.2 Oscillatory Cells and Cell Cultures

To describe the oscillatory activity of a cell, we modify the model by adding a constant depolarizing current to the ionic currents in (1). We model a two-dimensional cell culture by a square lattice with local diffusive coupling. This type of coupling represents electrical intercellular conductance coupling via the gap junctions.

The charge conservation equation for a lattice then reads

$$C_m \dot{v}_{ij} = -(I_{Na} + I_{si} + I_K + I_{K_1} + I_{K_p} + I_b) + I_{ij}^{stimulus} + D\Delta_d(v_{ij}) \quad (5)$$

where i, j are lattice indices, $I_{ij}^{stimulus}$ is the sum of two external currents: $I_{ij}^{stimulus} = I_{ij}^d + I_{ij}^a$, $I_{ij}^d > 0$ is a constant depolarizing current, I_{ij}^a is periodic in time current. Both external currents are non-identical in different cells. D is the coupling coefficient, and Δd is the second-order central difference operator (discrete Laplacian). A one-dimensional modification of this model is obtained by dropping the second spatial index.

Let us first study the single cell subject of only constant current I^d. In dependence on I^d the isolated cell can be oscillatory or excitable. When the value of I^d is increased above a bifurcation value approximate equal to 2.21 at the chosen values of parameters, a limit cycle appears in the phase space of the model, thus the cell becomes oscillatory. With further increase of I^d at $I^d = 4 : 05$ the cell becomes again excitable. So in the interval $I^d \in [2 : 21; 4 : 05]$ isolated cell demonstrate self-oscillations. Though this approach might not account for real physiological mechanisms of cell oscillation, the development of a more adequate model is hindered by the lack of understanding of the mentioned mechanisms in in-vitro experiments. However, in real situations, it is known that the leakage (depolarization) current of the non-pacemaker cells can increase turning them into oscillatory cells when they are dissociated from the heart tissues [14].

The measured dependence of the oscillation frequency of the cell upon the value of the depolarizing current I^d is presented in Fig. 1. Note that the spatial scale of one cell in the lattice model corresponds to the characteristic scale of culture inhomogeneity rather then to the size of a single cardiac cell.

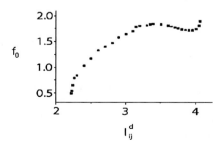

Fig. 1 Frequency of oscillations of an isolated Luo-Rudy cell versus the constant depolarizing current I^d. Outside the interval [2:21 ; 4:05] the cell is excitable

3 Effects of Mutual Synchronization in One-Dimensional Models

To get insight into some basic mechanisms, we start with a one-dimensional (1D) version of the model (5). We consider two different settings. In the first one the chain consists purely of oscillatory cells, and in the second one it is a mixture of oscillatory and excitable cells.

3.1 Ensembles of Oscillatory Cells

First, we study a chain of $N = 400$ oscillatory cells with different natural oscillation frequencies. For this we use quenched random depolarizing currents I_i^d, uniformly distributed in the interval [2:4; 3:2] $\mu A/cm^2$. We simulate the total of 10 chains with different realizations of this random distribution. The initial conditions are chosen to be identical in each cell, so that all cells in the chain initially get depolarized simultaneously. We simulate the system dynamics on the interval of 8×10^5 time units. Within this interval, we allow for a transient time of $T_{tr} = 4 \times 10^5$ units for the transient processes to be over and a stationary regime to set in. The duration of T_{tr} is chosen in a way that its further increasing does not lead to changes in the measurement results. In the subsequent observation time of $T_{ob} = 4 \times 10^5$ units we measure the individual average oscillation frequencies of each element. For that we define the section plane for the ith element and register each crossing of the trajectory with each of these section planes. We estimate the average oscillation frequency of the ith element as

$$v_i = v_s, \; \dot{v}_i > 0, \; v_s = -30.0,$$
$$f_i = (n_i - 1)/\Delta t_i \qquad (6)$$

where n_i is the number of crossings registered for the ith element, and Δt_i is the time elapsed between the first and the last crossing.

In Fig. 2(a) we plot the frequencies f_i of all elements in a chain with one realization of the random distribution of I_i^d versus D with dots. We see, that global synchronization sets in with increasing D, and the transition to global synchronization occurs via cluster regimes. A cluster regime is represented by a set of separated dots for a given value of D (say, $D = 0:006$). Each such dot corresponds to a frequency cluster. In Fig. 3(a–d) we plot the frequency profiles f_i versus element number i for several values of the coupling coefficient D in the same chain. We observe, that the size of clusters is gradually increasing, leading to a global synchronization regime (Fig. 3(d)), when all observed frequencies are equal up to the numerical estimation accuracy. In Fig. 4(a–d) we present the corresponding space-time color code plots of voltage in the chain, taken after the waiting time of 8×10^8 units. We observe a pacemaker (a local source of waves) in each cluster. A pacemaker is associated with a column of local minima of color lines on a space-time plot. In the global synchronous regime only one pacemaker remains. We observe a qualitatively

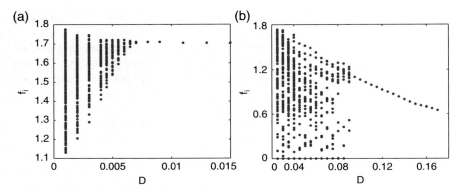

Fig. 2 Distribution of measured oscillation frequencies in a chain of $N = 400$ Luo-Rudy cells versus coupling coefficient D. Quenched random depolarizing currents I^d are distributed uniformly on the interval [2:4 ; 3:2] (**a**) and [0; 3:2] (**b**)

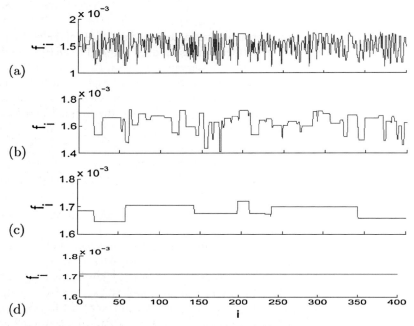

Fig. 3 Measured oscillation frequencies in the oscillatory chain versus cell number i at different values of the coupling coefficient $D = 0 : 001$ (**a**), 0.004 (**b**), 0.006 (**c**), 0.008 (**d**)

similar behavior for all 10 tested realizations of the random quenched depolarizing current. For a more detailed study of cluster synchronization in the system, we introduce its quantitative measure as the ratio of the maximal cluster size in the system N_c to the total system size N. Global synchronization regime thus corresponds to $N_c/N = 1$. We define a cluster as a set of adjacent cells with measured average

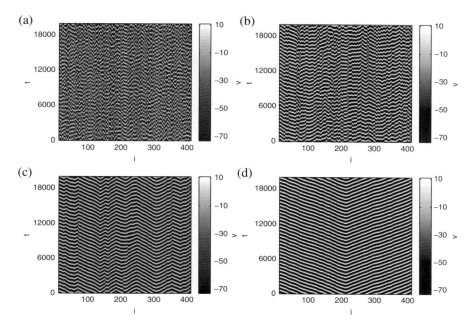

Fig. 4 Space-time plots of membrane voltage v in the oscillatory chain after waiting time 8×10^5 units at different values of the coupling coefficient $D = 0.001$ (**a**), 0.004 (**b**), 0.006 (**c**), 0.008 (**d**)

frequencies falling within the same error interval of size defined as $\Delta f = 2/T_{ob}$. In this measurement $\Delta f = 5 \times 10^{-6}$. We plot the ratio N_c/N for 10 realizations of the depolarizing current in Fig. 5(a). We observe the ratio generically growing with D, ultimately reaching the value 1. The points falling out of the bulk are due to the randomness in the simulations.

3.2 Mixtures of Oscillatory and Excitable Cells

Next, we consider a chain, which consists of a mixture of excitable and oscillatory Luo-Rudy cells. As heart tissue contains both types of cells, the problem of their interaction was actively studied [15–18]. To obtain a model of a mixture we change the interval of uniform distribution of the depolarizing currents to [0; 3:2]. From the numerically found value of the bifurcation point in 1 d we conclude, that about 31% of cells are oscillatory when uncoupled, and the other cells are excitable. We perform the same computational analysis of the model as in the previous setting. We plot average frequencies of all elements in a chain with one realization of the random distribution of $I^d{}_i$ versus D in Fig. 2(b). The only visible qualitative difference from the case of purely oscillatory chain is that the range of observed frequencies is now starting from zero. Note that the transition to global synchronization occurs at a higher value of D than in the oscillatory case. Next, we plot the frequency

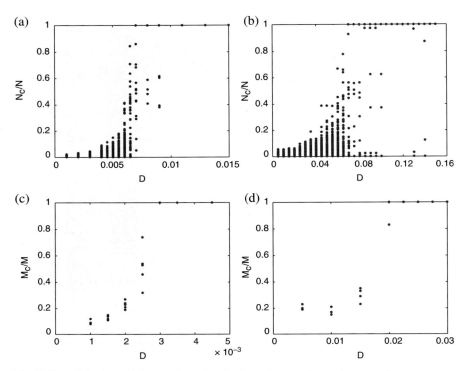

Fig. 5 Size of the largest cluster of synchronization related to the total system size versus coupling coefficient D:(a,b) – in chains of $N = 400$ cells for 10 different realizations of the uniform random distribution of I_i^d in the intervals [2:4; 3:2] (**a**) and [0; 3:2] (**b**); (**c, d**) – in lattices of $N = M \times M$ cells, $M = 100$, for 5 and 4 different realizations of the same two distributions, respectively

profiles f_i versus cell number i for several values of the coupling coefficient D in the same chain in Fig. 6(a–d). As expected, at small coupling the chain contains narrow groups of oscillating cells, separated by groups of cells at rest, which may be coined zero-frequency clusters (this means in fact, that the driving from neighboring cells is not enough for them to get membrane voltage above v_s). As coupling is increased, the non-zero frequency clusters are typically growing at the expense of zero-frequency ones. Note, that adjacent clusters with frequencies related as small natural numbers (like 1:2 or 2:3) are sometimes observed, see Fig. 6(b,c). This means, that the propagation of a certain fraction of the pulses (each 2nd or each 3rd in the mentioned examples) from the pacemaker into these regions is suppressed. Like in the case of purely oscillatory system, ultimately the regime of global synchronization sets in (see Fig. 6(d)). The ratio Nc = N for 10 realizations of the depolarizing current is plotted in Fig. 5(b). This ratio is generically growing with increasing D, reaching the value 1 at higher values of D, than in the case of purely oscillatory chain.

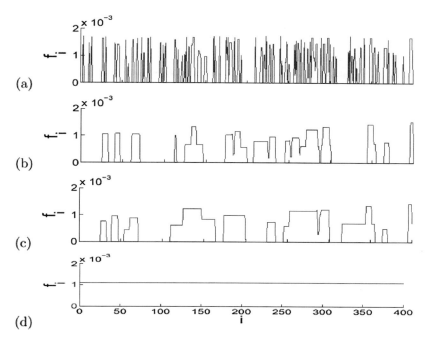

Fig. 6 Measured oscillation frequencies in the chain with oscillatory and excitable elements versus cell number i at different values of the coupling coefficient $D = 0.005$ (**a**), 0.03 (**b**), 0.04 (**c**), 0.075 (**d**)

We also plot the fraction Nz/N of non-excited (zero-frequency) cells for 10 realizations of I^d, see Fig. 7(a). As expected, this fraction is falling from about 0.7 down to zero. We carried out also simulations of a chain with the depolarizing currents distributed according to the Gaussian law with mean value equal to 2.8 and standard deviation equal to 0.5. The same qualitative results were reproduced.

The transition to global synchronization occurs around $D = 0.03$. In order to simulate the interaction between two pacemakers in the heart -sino-atrial and atrio-ventricular nodes – we consider the thirty elements chain of coupled cells: two oscillatory cells located at the ends of chain and excitable cells between them. The value of external currents of the first pacemaker was $I^d = 2.8$ and of the second one $I^d = 2.2$, thus these oscillatory cells had essentially different natural frequencies. Other elements had $I^d = 2$ and, hence, formed the excitable structure.

Figure 8 shows the dependence between cells frequencies and coupling coefficient D. We can see, that in the area of small D, excitable cells do not oscillate meanwhile the pacemakers interact slightly and almost do not impact each other. With growth of D frequencies of excitable cells start to move towards the frequency of the second pacemaker which has a less natural frequency than the first one. Starting from $D = 0.0056$ all excitable elements and the second pacemaker get into synchronous regime. Eventually they have the same frequency as the first pacemaker.

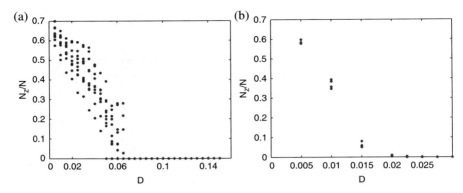

Fig. 7 Number N_z of non-excited elements related to the total system size N versus coupling coefficient D: (**a**) – in chains of $N = 400$ cells for 10 different realizations of the uniform random distribution of quenched depolarizing currents I^d_i in the interval $[0; 3:2]$; (**b**) – in lattices of $N = M \times M$ cells, $M = 100$, for 4 different realizations of the same distribution

Then we considered a chain with modified structure. Still it was a thirty-element chain but first twelve cells were oscillatory with $I^d = 2:8$, last twelve cells were also oscillatory with $I^d = 2:65$ and only six excitable cells with $I^d = 2$ were between them. The result of cells frequencies calculating is shown in Fig. 9(a). We can see that for $D = 0.003$–0.004 twelve pacemakers which initially had less natural

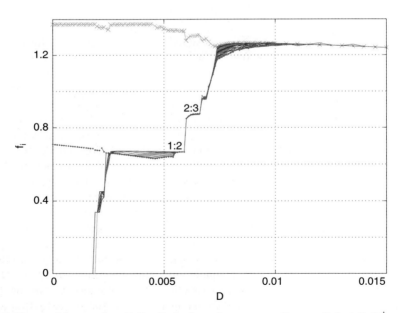

Fig. 8 Measured frequencies of all cells in dependence on coupling coefficient D. $I_1^d = 2:8$; $I_{30}^d = 2:2$; $I_{2-29}^d = 2$

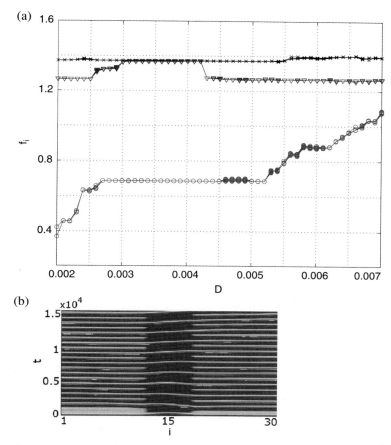

Fig. 9 (a) Measured frequencies of cells in dependence on coupling coefficient D. $I_{1-12}^d = 2:8$; $I_{19-30}^d = 2.65$; $I_{13-18}^d = 2$. (b) Space-time diagram of that process for $D = 0:004$

frequency get into synchronous regime with the other pacemakers. The important thing to notice is that all the excitable cells are synchronized with pacemakers with ratio 1:2. Figure 9(b) shows space-time diagram for this case

4 Two-Dimensional Models

In this section we show, that the main results obtained from the 1D models are kept in 2D models as well. We consider a square lattice of $N = M \times M$, $M = 100$, Luo-Rudy cells (5) with the same two distributions of quenched random depolarizing currents as in the 1D case. We perform simulations with 5 different realizations for the case of purely oscillatory system, and with 4 realizations for a mixture of oscillatory and excitable cells. The initial conditions are the same as in the 1D case. The total simulation time interval is 8×10^5 time units. As the transient processes appear to be longer in 2D case than in the 1D one, we choose transient time $T_{tr} = 6 \times 10^5$ units, and observation time $T_{ob} = 2 \times 10^5$ units. Similar to the 1D case, the transition to global synchronization occurs via cluster synchronization regimes. We measure the maximal cluster size Mc in horizontal and vertical directions in a way analogous to that taken in the 1D case. The frequency error interval is taken $\Delta f = 2/T_{ob} = 1 \times 10^{-5}$. The ratio $M_c = M$ is plotted in Fig. 5(c) and is qualitatively similar to that obtained in the 1D model. However, now $M_c = M = 1$ does not imply global synchronization, because local frequency defects are possible (see below).

Figure 10 shows the measured average oscillation frequency profiles for one realization of the quenched random current distribution on the interval [2.4; 3.2] at 4 different values of the coupling coefficient D along with corresponding snapshots of membrane voltage v_{ij} in the end of 8×10^5 units time interval. At small coupling $D = 0.001$, frequency clusters are formed, but consist of no more than a few cells, and the activity in the lattice looks incoherent (Fig. 10(a,b)). As coupling D is increased, the clusters get larger (Fig. 10(c,d)). After further increasing D, almost the whole lattice gets covered with one cluster, except for small "defects" characterized by differing frequencies (Fig. 10(e,f)). The corresponding space-time evolution in the latter case is an almost regular target wave structure, but it contains defects in the forms of additional pacemakers and spiral cores, which can coexist (Fig. 10(f)). Such structural defects and the mentioned defects in the frequency profiles are typically well associated with each other (compare Fig. 10(e) and (f)). Further increasing the coupling parameter leads to a globally synchronous regime. We observe, that it can be represented as well by one pacemaker, two pacemakers and a spiral wave in different realizations of 1D distribution (Fig. 11(a,b,c), respectively). However, it is impossible to determine with computational methods, whether two pacemakers indeed do coexist and are frequency-locked (as for 1D case reported above), or the finite transient time is insufficient to observe one of them being destroyed, and the observation time in not enough to resolve their frequency difference. The ratio Mc = M for 4 realizations of the mixture of oscillatory and excitable cells is plotted versus D in Fig. 5(d), the ratio of the number of non-excited elements to the total number of elements is presented in Fig. 7(b). We, observe, that with increasing D frequency clusters are growing, and the fraction of never excited elements is falling to zero. The space-time evolution is characterized by a transition from spatially-incoherent behavior to globally synchronous regimes driven by pacemakers or spirals.

5 Phenomena of External Synchronization

In this section we study the influence of external force on the collective dynamics in 1D and 2D networks. Our experiments based on the fundamental property, that in excitable media the wave with the highest frequency will overtake all other waves.

Synchronization Phenomena in Networks

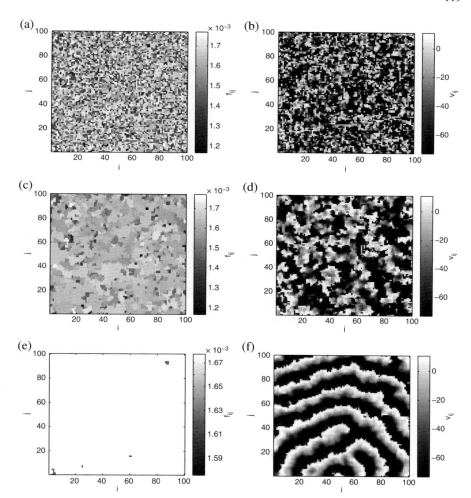

Fig. 10 Measured oscillation frequencies (**a, c, e**) and snapshots of membrane voltage after waiting time 8×10^5 units (**b, d, f**) in a 2D lattice of 100×100 oscillatory Luo-Rudy cells at different values of the coupling coefficient $D = 0.001$ (**a, b**), 0.002 (**c, d**), $D = 0.003$ (**e, f**). Quenched random depolarizing currents I_{ij}^d are distributed uniformly on the interval [2.4; 3.2]

The task was to find the high-frequency response of the cells for the external pacing. We add a current I_{ij}^{stimulus} in Eq. (5), where $i; j$ are the lattice indices, I_{ij}^{stimulus} is the sum of two external currents: I_{ij}^d is constant depolarizing current, I_{ij}^a is periodic in time current.

$$I_{ij}^{\text{stimulus}} = I_{ij}^d + I_{ij}^a \qquad (7)$$

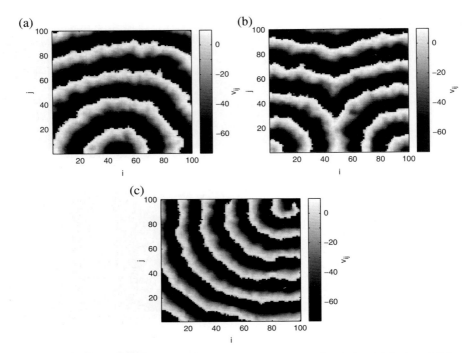

Fig. 11 Snapshots of membrane voltage after waiting time 8×10^5 units in a 2D lattice of 100×100 oscillatory Luo-Rudy cells for three different realizations of the uniform random distribution of quenched depolarizing currents I_{ij}^d at $D = 0.004$. In all cases global synchronization up to numerical accuracy is observed

In our first series of these computational experiments, we periodically paced one isolated Luo-Rudy cell by the square-wave stimulus – 50% duty cycle, amplitude of I_{ij}^a:$A = 2, 5, 10\,\text{mA/cm}^2$. We measure the output frequency that depends on the frequency of pacing. Different types of synchronization phenomena take place for different values of input currents. Figure 12 shows the examples of 1:1 and 2:1 synchrony. It was found that:

i) The pacing with the amplitude of the input low-amplitude high-frequency current (50% duty cycle) $A = 2\,\text{mA/cm}^2$ don't allows to get a high-frequency response of an excitable Luo-Rudy cell. We can get the highest output frequency for an oscillatory cell. The output frequency may equals to 15.7 Hz for the input current's frequency equals 32.5 Hz (synchronization 2:1 takes place).

ii) If the amplitude of the input signal is $A = 5\,\text{mA/cm}^2$, synchronization 1:1, 2:1, 3:1 is observed. If the low-amplitude constant current is higher than $I^d = 3\,\text{mA/cm}^2$, the output frequency equal 65–80 Hz (synchronization 1:1) can be found.

Synchronization Phenomena in Networks 121

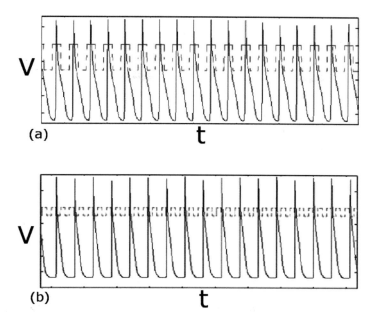

Fig. 12 Different types of synchronization (**a**) 1:1, (**b**) 2:1 for the frequency of the pacing and the output frequency. *Dotted line* shows the input signal in diagram form, firm line shows the action potential

iii) If the amplitude of the input high-frequency signal is $A = 10\,\text{mA/cm}^2$, synchronization 1:1 takes place for all values of the low-amplitude constant current.

The results are presented in Fig. 13. We state maximal values of the output frequencies of the cell versus input values of the low-amplitude constant current. One can see, that the most perspective amplitudes of the input low-amplitude constant current are:

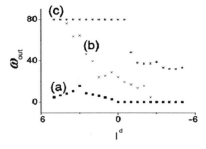

Fig. 13 Maximal output frequencies of one Luo-Rudy cell versus the input low-amplitude constant current. The amplitude of *square-wave* low-amplitude high-frequency stimulus (50% duty cycle) is (**a**) $A = 2\,\text{mA/cm}^2$; (**b**) $A = 5\,\text{mA/cm}^2$; (**c**) $A = 10\,\text{mA/cm}^2$

Fig. 14 APD (Action Potential Duration) versus the low-amplitude constant current I^d

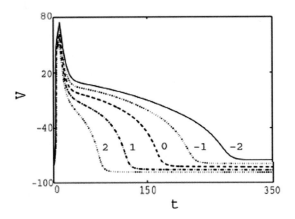

(i) for amplitude of $I^a_{ij}: A = 2\,\text{mA/cm}^2$ it is the area, which corresponds to oscillatory cell;
(ii) for $A = 5\,\text{mA/cm}^2$ and $A = 10\,\text{mA/cm}^2$ it may be the area of negative values of the input low-amplitude constant current. We investigate the influence of low-amplitude constant current to the APD (Action Potential Duration) of the Luo-Rudy cell. It is possible to say, that the application of I d leads to decrease APD (Fig. 14)

5.1 Synchronous Response in the Chain of Coupled Cells

We periodically paced the first cell in the chain of 10 Luo-Rudy cells with a square-wave stimulus – 50% duty cycle (amplitudes of the input high-frequency current are $A = 10; 30; 50\,\text{mA/cm}^2$). We measured the output frequency in the last element in the chain versus the frequency of external pulses. The results are presented in Fig. 15. Synchronization 1:1, 2:1, 3:1 and others take place.

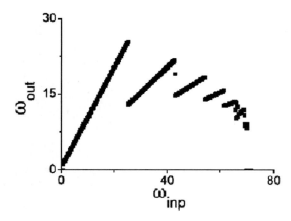

Fig. 15 The output frequency of oscillations of 10 Luo-Rudy cells (from the 10-th cell) versus the frequency of pacing. We paced only the first cell in the chain. The amplitude of the input high-frequency stimulus is $A = 30\,\text{mA/cm}^2$ (50% duty cycle), constant current is $I^d_{ij}: A = 2\,\text{mA/cm}^2$

Synchronization Phenomena in Networks

i) The amplitude of the input high-frequency current $A = 10\,\text{mA/cm}^2$ for 10 cells is insufficient for getting a high-frequency response of the last cell in the chain. Maximal frequency may be equal to 3.0764 Hz.

ii) Values of the input high-frequency current $A = 30\,\text{mA/cm}^2$ and $A = 50\,\text{mA/cm}^2$ allows to get frequencies 20–35 Hz (synchronization 1:1).

The maximal output frequencies versus the input values of the low-amplitude constant current I_{ij}^d are presented the in Fig. 16. We can establish a fact, that the most perspective amplitudes of the low-amplitude constant current are: (i) for the minimal amplitude of the input high-frequency current $A = 10\,\text{mA/cm}^2$ it is the area of values of constant current from $0.5\,\text{mA/cm}^2$ to $1.5\,\text{mA/cm}^2$; (ii) for $A = 30\,\text{mA/cm}^2$ it

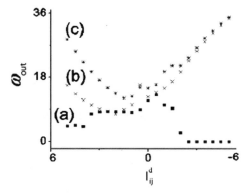

Fig. 16 Maximal output frequencies versus the input low-amplitude current for the chain of 10 Luo-Rudy cells. We paced only the first cell in the chain with the square-wave high-frequency stimulus (50% duty cycle). The amplitude of stimulus is (**a**) $A = 10\,\text{mA/cm}^2$; (**b**) $A = 30\,\text{mA/cm}^2$; (**c**) $A = 50\,\text{mA/cm}^2$

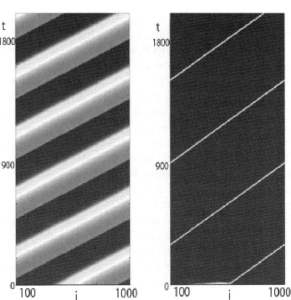

Fig. 17 Propagation of the pulse in the chain of 1000 elements Luo-Rudy for $I_{ij}^d = 2.4\,\text{mA/cm}^2$ (to the *left*) and $I_{ij}^d = -5\,\text{mA/cm}^2$ (to the *right*). The time-axis is in vertical direction, i is a number of element

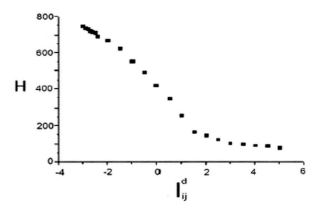

Fig. 18 The pulse width in the chain of 1000 Luo-Rudy elements versus the input low-amplitude constant current

is values of constant current less than $2\,\text{mA/cm}^2$; (iii) for $A = 50\,\text{mA/cm}^2$ it is values of the constant current less than $4\,\text{mA/cm}^2$. Because of individual cell dynamics the possibility of bistable synchronous response was observed. We investigate the propagation of the pulse in the chain of 1000 Luo-Rudy cells for different values of the low-amplitude constant current (Fig. 17). It was shown that the application of the positive current leads to decrease of the pulse width, if initial conditions are equivalent (Fig. 18). It is in agreement with the previous experiment for one cell.

6 Discussions

We have studied the dynamics of autonomous one- and two-dimensional inhomogeneous cardiac culture models in two settings: (i) an ensemble of oscillatory cells with different natural frequencies and (ii) a mixture of excitable and oscillatory cells. In all these kinds of models we observed the transition from incoherent behavior to global synchronization via cluster synchronization regimes when the coupling strength is increased. We have measured main quantitative characteristics of these regimes (distributions of local average frequencies and cluster sizes) in dependence upon the coupling strength. In two-dimensional models we observed various spatiotemporal patterns of activity, including target and spiral waves and complicated irregular behavior. From the consideration above it is clear that the coupling constant D plays an important role in the collective dynamics of the cells. In real situations, this D corresponds probably to the gap-junction connectivity between cells. In an experiment with cardiac cell cultures from chicken embryos, Glass and his coworkers [19] have recently shown that the control of connectivity of the system through the use of glycerrhetinic acid and cultures density can indeed produce a transition to synchronized patterns [19]. They have used a heterogeneous cellular automaton model to understand their experiment findings. It seems that heterogeneity and excitability are essential in their explanations. In our case, the system is oscillatory or is a mixture of excitable and oscillatory cells. Although cells taken from the ventricle [8] are considered to be only excitable when they are in an intact heart, they will

Synchronization Phenomena in Networks

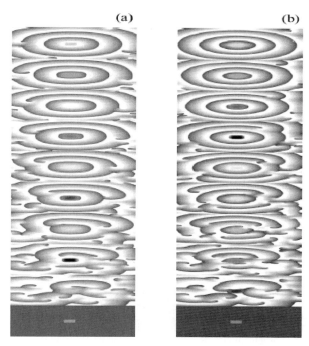

Fig. 19 Suppressing spiral waves chaos. Grid 300 × 300 Luo-Rudy cells. $A = 30\,\text{mA/cm}^2$, frequencies are (**a**) $\omega = 0{:}15$; (**b**) $\omega = 0{:}16$. Time axis is in vertical direction. The *first bar* shows the input signal

become oscillatory [14] after they have been dissociated from the heart tissue and plated on the culture dishes. It is known that their oscillation period will be much longer than that of the pacemakers, but it is not clear what the oscillation period distribution is and whether this distribution will depend on the growth conditions. Even though all the cells might be eventually oscillatory, as a first approximation, one can still consider most of the cells as excitable as they will be driven by a few cells with the shortest oscillation periods. It is therefore reasonable to assume that there is a mixture of excitable and oscillatory cells. In this sense, we also have heterogeneity and excitability built into our systems. However, the question is open whether this heterogeneity is the same as that in Ref. [19]. In non-autonomous case the results of our simulations suggest that it may be possible to defibrillate cardiac tissue using low-amplitude, high-frequency pacing in the presence of low-amplitude constant current. Latter leads to more regular behavior. Then by using the proper frequencies, amplitudes, and waveforms predicted from the one-dimensional simulations, we showed that it is possible to suppress spiral waves and spiral wave chaos in two-dimensional media (Fig. 19). A clear next step is to confirm these results experimentally in animal models.

Acknowledgments This research was supported by RFBR-NSC (Project No. 05-02-90567), RFBR-MFC (Project No. 05-02-19815), RFBR (Project Nos. 06-02-16596, 06-02-16499) and program "Leading scientific schools of Russia" (grant No. 7309.2006.2). O.K. also acknowledges support from the "Dynasty" Foundation, Russia, and J.K. that of the International Promotionskolleg Cognitive Neuroscience and BIOSIM.

References

1. Krinsky V.I. Spread of excitation in an inhomogeneous medium. Biophysica (USSR) 1966, 11:776–784.
2. Stamp A.T., Osipov G.V., and Collins J.J. Suppressing arrhythmias in cardiac models using overdrive pacing and calcium channel blockers. Chaos 2002, 12:931–940.
3. Weiss J.N., Qu Z., Chen P.S., Lin S.F., Karagueuzian H.S., Hayashi H., Garfinkel A., Karma A. The dynamics of cardiac fibrillation. Circulation 2005, 112:1232–1240.
4. Zhang H., Cao Zh., Wu N.-J., Ying H.-P., Hu G. Suppress Winfree turbulence by local forcing excitable systems. Phys. Rev. Lett. 2005, 94:188301.
5. Winfree A.T. (ed.), Focus issue: Fibrillation in normal ventricular myocardium. Chaos 1998, 8(1):1–241.
6. Christini D.J. and Glass L. (eds.), Focus issue: Mapping and control of complex cardiac arrhythmias. Chaos 2002, 12(3):732–981.
7. Kurths J., Wessel N., Bavernschmitt R., and Ditto W. (eds.), Focus issue: Cardiovascular physics. Chaos 2007, 17(1):015101–015121.
8. Hwang S.-M., Yea K.-H., and Lee K.J. Regular and alternant spiral waves of contractile motion on rat ventricle cell cultures. Phys. Rev. Lett. 2004, 92:198103.
9. Luo C.-H. and Rudy Y. A model of the ventricular cardiac action potential: Depolarization, repolarization, and their interaction. Circ. Res. 1991, 68(6):1501–1526.
10. Luo C.-H. and Rudy Y. A dynamic model of the cardiac ventricular action potential. I. Simulations of ionic currents and concentration changes. Circ. Res. 1994, 74:1071–1096.
11. Courtemanche M., Ramirez R.J., and Nattel S. Ionic mechanisms underlying human atrial action potential properties: Insights from a mathematical model. Am. J. Physiol. 1998, 275:H301H321.
12. Ten Tusscher K.H., Noble D., Noble P.J. Panfilov A.V. A model for human ventricular tissue. Am. J. Physiol. Heart Circ. Physiol. 2004, 286:H1573–H1589.
13. Ten Tusscher K.H. and Panfilov A.V. Alternans and spiral breakup in a human ventricular tissue model. Am. J. Physiol. Heart Circ. Physiol. 2006, 291(3): H1088–H1100.
14. Lilly S.L. "Pathophysiology of Heart Disease," 3rd ed., 2003, Lippincott Williams and Wilkins, USA.
15. Fozzard H.A. and Schoenberg M. Strengthduration curves in cardiac Purkinje fibres: Effects of liminal length and charge distribution. J. Physiol. 1972, 226(3):593–618.
16. Winslow R.L., Varghese A., Noble D., Adlakha C., and Hoythya A. Generation and propagation of ectopic beats induced by spatially localized Na-K pump inhibition in atrial network models. Proc. R. Soc. B 1993, 254:55–61.
17. Joyner R.W., Wang Y.G., Wilders R., Golod D.A., Wagner M.B., Kumar R., and Goolsby W.N. A spontaneously active focus drives a model atrial sheet more easily than a model ventricular sheet. Am. J. Physiol. 2000, 279:H752–H763.
18. Wilders R., Wagner M.B., Golod D.A., Kumar R., Wang Y.G., Goolsby W.N., Joyner R.W. and Jongsma H.J. Effects of anisotropy on the development of cardiac arrhythmias associated with focal activity. Pfl¨ugers Arch. 2000, 441:301–302.
19. Bub G., Shrier A., and Glass L. Global organization of dynamics in oscillatory heterogeneous excitable media. Phys. Rev. Lett. 2005, 94(2):028105.

Nonlinear Oscillations in the Conduction System of the Heart – A Model

Krzysztof Grudziński, Jan J. Żebrowski and Rafał Baranowski

Abstract The effects of the interaction of the phase space trajectory of a modified van der Pol oscillator with a hyperbolic saddle are discussed. It is shown that the refractory period is obtained and that the saddle affects the way the model reacts both to parameter change and to external perturbation. We obtain results comparable to effects observed in recordings of heart rate variability.

Keywords Nonlinear oscillations · Heart rate variability models

1 Introduction

Progress in the last decade made in the modeling of ion channels in cell membranes has lead to the development of well developed models of a number of biological cells such as neurons and myocytes [1]. Such models allow obtaining realistic properties of the action potentials generated by these cells which are comparable with experimental results. In recent years, many important results have been obtained from this class of models including those related to new drugs. Paradoxically, because of the progress in the understanding of the processes in the ion channels and due to the identification of a large number of these channels in the cell membranes, the ion channel models usually have a rather large parameter space and may be difficult to analyze. In addition, the detailed properties of the various ion channels need to be determined in separate experiments and many properties are still to be uncovered.

Recently, we developed a model of the conduction system of the heart – based on coupled nonlinear oscillators and not on the properties of ion channels [2–4]. The conducting system of the human heart: the SA node (the primary pacemaker), the AV node and the His-Purkinje system – may be treated as a network of self excitatory elements. These elements may be modelled as interacting nonlinear oscillators [5]. The phenomenological approach using nonlinear oscillators is not suitable

---

K. Grudziński (✉)
Physics of Complex Systems, Warsaw University of Technology,
ul.Koszykowa 75, Warszawa, Poland

for the investigation of the cardiac conducting system at a cellular level. It allows, however, a global analysis of heartbeat dynamics by investigating interactions between the elements of the system. We were interested in finding a simple model in order to ascertain what is the minimal structure of the phase space that allows the model to perform similarly to the conduction system of the heart. In the past, nonlinear oscillator models of heartbeat dynamics (see the paper by di Bernardo [6] and references therein) assumed bi-directional coupling between the SA and AV nodes. Such a coupling signifies that the *SA node reacts immediately* to the dynamics of the AV node. In our model of the conduction system [2,3], we assumed a uni-directional coupling between the nodes.

The reason why we chose relaxation oscillators for our model of the conduction system of the heart was twofold. The van der Pol oscillator adapts – without changing the amplitude [7] – its intrinsic frequency to the frequency of the external driving signal $F(t)$. This is a very important feature because the main cardiac pacemaker is the element of the conducting system with the highest frequency to which all other oscillators must adjust. As a result, a hierarchy of pacemakers in the conduction system of the heart is created – an observation which led van der Pol and van der Mark to the first model of the heart [8] and to the famous van der Pol relaxation oscillator model. The second reason was that FitzHugh [9] showed how an extended version of the van der Pol model may be obtained as a simplification of the Hodgkin-Huxley equations. The FitzHugh-Nagumo model [1,9] may be treated as a link between relaxation oscillator models and modern analysis of physiological oscillators based on ion channels. Currently, both FitzHugh-Nagumo and Hodgkin-Huxley type models are used to study the properties of cardiac tissue [1].

Several other groups have published results also using models based on relaxation oscillators [10–12]. The research was focused on the interaction between the AV and SA nodes. Although many interesting results obtained, some important features of real cardiac action potentials were not reproduced. The most important were the shape, the properties of the refractory period as well as the modes of pulse frequency change. These properties of real action potentials are today reproduced by ion channel models such as those discussed in Ref. [1, 13].

We found that the way to obtain these features was to introduce into the phase space of the system a hyperbolic saddle and a node besides the unstable focus that occurs in the phase space of the van der Pol model (Fig. 1). Whatever the values of the parameters of the model, the node, the saddle and the unstable focus always appear in that order. The nonlinear oscillator we developed is based on the work by Postnov et al. [14].

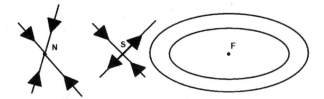

Fig. 1 Sketch of the phase space of the nonlinear oscillator used in our work.
N – stable node,
S – hyperbolic saddle,
F – unstable focus

The aim of this paper is to study the effect of the structures introduced into the phase space of the modified van der Pol system in Refs. [2–4]. In particular, we show how the refraction period (i.e. the part of the oscillator period during which it is insensitive to external perturbations) occurs due to the combined effect of the properties of the stable node and of the saddle on the trajectory of the system. We also discuss the change of the period of oscillations due to a single external pulse and the possibility of a cessation of oscillations (asystole) such a pulse may introduce.

2 The Model

In their paper on synchronization of nonlinear oscillators, Postnov et al. [14] developed a modified van der Pol model. For our purposes, their model did not allow a convenient way of changing the frequency of the oscillations. We needed to add two new parameters into the Duffing term of their equation [3, 4]. In addition, we modified the relaxation term also and obtained an oscillator [3] that reproduces two of the three ways that a node of the conduction system of the heart may change its frequency [15]. The model equation we obtained [2–4] is:

$$\frac{d^2x}{dt^2} + \alpha(x-v_1)(x-v_2)\frac{dx}{dt} + f\,x(x+d)(x+e) = 0 \qquad (1)$$

where $x(t)$ is the potential and all structure in phase space is situated along the horizontal axis so that $x = 0$ is position of the unstable focus, $x = -d$ is the position of the hyperbolic saddle and $x = -e$ is the position of the stable node. The parameter f allows to rescale the frequency of the oscillations [3]. α is the damping and $v_1 v_2 < 0$ to allow self-oscillations.

In our calculations, we set $\alpha = 5$, $d = 3$, $v_1 = 1$, $v_2 = -1$, and $f = 3$. To maintain the proper ratio of the frequencies of the two nodes, to model the SA node we used $e = 12$ and to model the AV node we set $e = 7$ [2–4]. All variables and parameters in this paper are dimensionless but the magnitude of the period of the oscillations was kept in a range similar to that observed in physiology.

3 Results and Discussion

Since we would like to construct a model of the conduction system of the heart, we need to place the different parameters of our model within a physiological context. It is well known that an increase of rate heart due to the activity of the sympathetic nervous system has an upper bound. In addition, as sympathetic activity increases the variability of the heart decreases. Figure 2 depicts the change of the period of the oscillator as a function of parameter e [2]. It can be seen that the period is a decreasing function of this parameter and that as e increases the curve saturates. The saturation is in keeping with the physiological property of a limited maximum

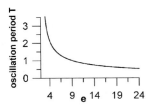

Fig. 2 Saturation of the period of oscillations with an increase of the parameter e. This parameter controls the position of the stable node

heart rate. In addition, because of the saturation the variability of the period due to fluctuations in the value of the parameter e will also decrease. Thus, increasing the absolute value of the parameter e of Eq. (1) is equivalent to an increase of sympathetic activity.

The parasympathetic nervous system acts on the conduction system of the heart through such neurotransmitters as acetylocholine. This neurotransmitter has a very fast rise time and quickly decays. In Ref. [3] we modeled the effect of acetylocholine using short, negative spikes with noise frequency. As a first approximation, however, we may also model the effect of acetylocholine by applying a short rectangular pulse of potential to our model. Figure 3 depicts the effect of such a pulse with amplitude $A = -2$ applied during the period of vulnerability of the oscillator. It can be seen that the next pulse of the action potential is delayed i.e. the instantaneous period of the oscillation was decreased. Note how the saddle situated at $(-3, 0)$ affects the shape of the perturbed trajectory (thin line in Fig. 3).

The exact effect the external pulse has on the oscillation depends on the phase at which that pulse is applied. Figure 4 depicts the phase response curve for the oscillator Eq. (1) for the amplitude of the applied pulse $A = -2$ and $A = -1$ and a constant width of $L = 0.05$. The zero of the phase was set at the maximum of the action potential. The phase change at negative applied phase indicates that the application of the external pulse affected also the next pulse generated by the oscillator. When the external pulse was applied at the end of the refractory phase (Fig. 5), the resultant instantaneous period of oscillations was shortened. If the area under the curve of the applied external pulse is chosen properly a long pause lasting several normal periods of the oscillation may occur – similarly as it may occur in the human heart due to a single supraventricular ectopic beat [3]. The nonlinearity

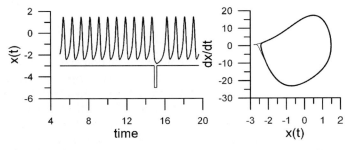

Fig. 3 Effect of the application of a single pulse: amplitude $A = -2$, pulse width $L = 0.25$. The pulse was applied at $t = 14.9$ which is equivalent to the phase $\varphi = 0.24$. *Left Panel*: Potential as a function of the time t; *Right Panel*: phase portrait; the *thin line* depicts the perturbed trajectory

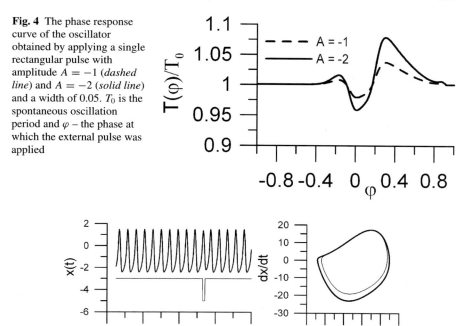

Fig. 4 The phase response curve of the oscillator obtained by applying a single rectangular pulse with amplitude $A = -1$ (*dashed line*) and $A = -2$ (*solid line*) and a width of 0.05. T_0 is the spontaneous oscillation period and φ – the phase at which the external pulse was applied

Fig. 5 Effect of the application of a single pulse: amplitude $A = -2$, pulse width $L = 0.25$. The pulse was applied at $t = 14.6$ equivalent to the phase $\varphi = 0.93$. *Left Panel*: Potential as a function of the time; *Right Panel*: phase portrait; the *thin line* depicts the perturbed trajectory

of the phase response curve results in an irregular behaviour of the oscillations when an external periodic pulse train is applied [3].

Positive external single pulses usually do not have such a strong effect on the instantaneous period of the oscillator. This is due to the symmetry breaking effect of the hyperbolic saddle situated on one side of the limit cycle only. Two examples of the effect of single positive pulses applied at the end of the refractory period and during the vulnerability period are shown in Figs. 6 and 7, respectively. It can be seen that, in the first case, the period is shortened slightly because the trajectory was deflected away from the saddle and towards the center of the limit cycle (thin line in Fig. 6). By this, the trajectory avoided the slowing down near the unstable fixed point. In the latter case, the effect is stronger and due to two combined effects. The period is longer than normal because the perturbation decreases temporarily the resting potential (Fig. 7 Left Panel) and also the trajectory is deflected towards the saddle (Fig. 7 Right Panel) so that the critical slowing down has a larger effect.

A large amplitude positive pulse may induce a change of period affecting two consecutive oscillations. An example of such a behavior is depicted in Fig. 8, where a pulse with amplitude $A = 6$, width $L = 0.25$ was applied at $\varphi = 0.24$. This is similar to the effect an ectopic beat often has: it is well known that an arrhythmia may perturb not only the RR interval during which it occurs but also the subsequent one.

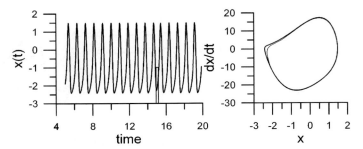

Fig. 6 Effect of the application of a single pulse: amplitude $A = 2$, pulse width $L = 0.25$. The pulse was applied at $\varphi = 0.24$. *Left Panel*: Potential as a function of the time; *Right Panel*: phase portrait; the thin line depicts the perturbed trajectory

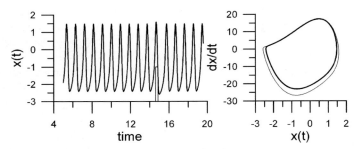

Fig. 7 Effect of the application of a single pulse: amplitude $A = 2$, pulse width $L = 0.25$. The pulse was applied at $\varphi = 0.93$. *Left Panel*: Potential as a function of the time; *Right Panel*: phase portrait; the thin line depicts the perturbed trajectory

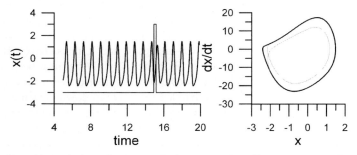

Fig. 8 Large positive pulse (amplitude $A = 6$, phase $\varphi = 0.24$) affects two consecutive periods of the oscillation. The *thin line* depicts the perturbed trajectory

Nonlinear Oscillations in the Conduction System of the Heart 133

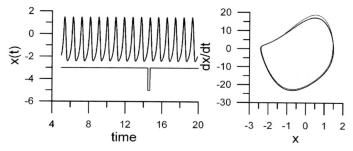

Fig. 9 Insensitivity of the trajectory to a negative external applied during the refractory period. The pulse amplitude $A = -2$, width $L = 0.25$ and phase $\varphi = 0.77$. The *thin line* depicts the perturbed trajectory

During the refractory period, neither a negative external pulse (Fig. 9) nor a positive one (Fig. 10) results in a change of the period. The perturbed trajectory is able to return to the limit cycle in such a way that the passage close to the saddle is unaffected. During this passage, the phase of the oscillation equivalent to spontaneous depolarization of the real action potential occurs [3]. A relatively large part of the oscillation period is spent by the system close to the saddle.

The effect of several external pulses may cumulate. This property of our oscillator is important because usually the emission of acetylcholine occurs in the form of bursts of many spikes (see discussion of the vagal paradox in Ref. [3]). Figure 11 depicts the effect of the application of a rectangular wave with a relatively large frequency. It can be seen that some of the pulses have been substituted by subthreshold oscillations. As a result the effective period of the oscillations has been substantially extended. The number of subthreshold oscillations between the normal pulses of the oscillator may be controlled by adjusting the amplitude, length and frequency of the rectangular perturbation. This property of our model allows reproducing not only the effect of acetylocholine bursts but also that of a fast parasystole. Note that the amplitude of the perturbing rectangular wave may not exceed the critical value at which the trajectory of the oscillator will be flipped across the saddle onto the stable node and the oscillations will cease (asystole, see [2, 3]).

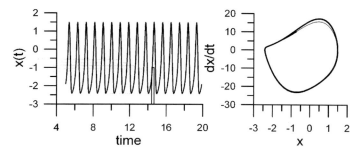

Fig. 10 Insensitivity of the trajectory to a positive external applied during the refractory period. The pulse amplitude $A = 2$, width $L = 0.25$ and phase $\varphi = 0.72$. The *thin line* depicts the perturbed trajectory

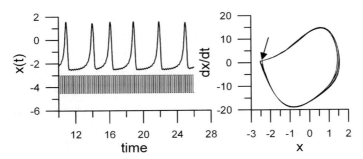

Fig. 11 A fast *rectangular wave* ($A = -1.5$, $L = 0.1$, frequency 0.15) applied to the oscillator results in subthreshold oscillations and a decrease of the average period of the oscillations. During subthreshold oscillations the state points resides in the vicinity indicated by the arrow

A rectangular wave with a frequency smaller than that of the oscillator modifies the individual phases of the oscillator pulses. An asystole may then also occur just as for a single external pulse applied at the right phase and with the correct amplitude and length [2, 3]. However, for intermediate frequencies of the external square wave perturbation irregular behavior of the oscillator may result [3].

4 Summary and Conclusions

In a series of recent publications [2–4], we have discussed the possibility of modeling the properties of the conduction system of the heart using a modified van der Pol oscillator as a building block. We treat the His-Purkinje system as a passive delay and since, at this stage of our research, we do not discuss the effect of feedback – we simply ignore this part of the conduction system. The two remaining elements of the conduction system – the SA and the AV nodes we model as unidirectionally and diffusively coupled oscillators setting the ratio of their frequencies accordingly. We showed that the minimum phase space configuration that enables to reproduce several properties of the natural pacemakers of the heart consists of a stable node, an unstable saddle and an unstable focus. An oscillator with such a phase space reproduces many of the properties of the natural pacemakers hitherto obtained in the much more complex ion channel models. The properties reproduced are: the shape of the oscillation pulse, the refratory period and the phase sensitivity [2–4]. Our model changes the frequency of its oscillations using two most frequently encountered of the three possible modes of frequency change of a natural pacemaker [2, 3]. Other features of our model which can be directly compared with phenomena measured in clinical cardiology or physiology include [2, 3]:

- saturation of the heart rate with an increase of sympathetic activity and the related decrease of heart rate variability,
- sinus node pause,

- asystole of the node due to an external perturbation (such as e.g. an ectopic beat; it is possible to switch the rhythm back on),
- the vagal paradox (hitherto obtained in animal experiments [16] and by means of an ion channel model of the pacemaker cell [17])
- unusual patterns in the heart rhythm associated with heart block.

In addition, our model predicts effects which may or may not be observable [2,3]:

- the effect of the asymmetry of the SA-AV diffusive coupling which may lead to an amplitude death of the AV node
- type I intermittency in the interspike interval dynamics due a parametric modulation of the properties of the conduction system node properties
- the cessation of oscillations ending in a deeply depolarized node (i.e. the oscillator trajectory rests at the stable fixed point)
- regular pacing of the single node results in an irregular response.

Most of the effects summarized above are due to introducing into the phase space of the system the stable node and the unstable saddle. In this paper, we focused on the interaction of the trajectory of the oscillator with the saddle. By numerical experiments, we showed how the refractory period is a direct result of the interaction of the trajectory with the stable and unstable manifolds of the saddle. The properties of these manifolds were modified by shifting the position of the stable node. The existence of the hyperbolic saddle in the phase space results also in subthreshold oscillations due to the application of a square wave perturbation with a relatively high frequency. This is a model of the effect of acetylocholine – the neurotransmitter which is associated with the vagal system and is often released in high frequency bursts. The basic action of the vagal system is to slow down the heart rate and that is what our model shows. Finally, we found that a strong external pulse – such as an action potential produced by an ectopic source – may modify not only the phase of the next pulse of our oscillator but also that of the next one. Such a phenomenon is commonly obtained in ECG recordings.

Acknowledgments This paper as a project of the ESF Programme "Stochastic Dynamics: fundamentals and applications (STOCHDYN)" was supported by Polish Ministry of Science and Higher Education, Grant No. ESF/275/2006. The authors gratefully acknowledge very stimulating discussions with Teodor Buchner and Paweł Kuklik.

References

1. J. Keener, J. Sneyd, *Mathematical Physiology*, Interdisciplinary Applied Mathematics **8** (Springer, New York, 1998).
2. K. Grudzinski, J.J. Żebrowski, R. Baranowski, "A model of the sino-atrial and the atrio-ventricular nodes of the conduction system of the human heart", Biomed. Eng. **51**, 210–214 (2006).
3. J.J. Żebrowski, K. Grudziński, T. Buchner, P. Kuklik, J. Gac, G. Gielerak, P. Sanders, R. Baranowski, "A nonlinear oscillator model reproducing various phenomena in the dynamics

of the conduction system of the heart", to appear in Chaos **17**, Focus Issue "Cardiovascular Physics", (2007).
4. K. Grudziński, J.J. Żebrowski, "Modeling cardiac pacemakers with relaxation oscillators", Physica A **336**, 153–152 (2004).
5. G. Bub, L. Glass, "Bifurcations in a discontinuous circle map: a theory for a chaotic cardiac arrhythmia", Int. J. Bifurcations and Chaos **5**, 359–371 (1995).
6. D. di Bernardo, M.G. Signorini, S. Cerutti, "A model of two nonlinear coupled oscillators for the study of heartbeat dynamics", Int. J. Bifurcations and Chaos **8**, 1975–1985 (1998).
7. J.M.T. Thompson, H.B. Steward, *Nonlinear Dynamics and Chaos* (Wiley, New York, 2002).
8. B. van der Pol, J. van der Mark, "The heartbeat considered as a relaxation oscillation and an electrical model of the heart", Phil. Mag. **6**, 763–775 (1928).
9. R. FitzHugh, "Impulses and physiological states in theoretical models of nerve membrane", Biophys. J. **1**, 445–466 (1961).
10. C.R. Katholi, F. Urthaler, J. Macy Jr., T.N. James, "A mathematical model of automaticity in the sinus node and the AV junction based on weakly coupled relaxation oscillators", Comp. Biomed. Res. **10**, 529–543 (1977).
11. J. Honerkamp, "The heart as a system of coupled nonlinear oscillators", J. Math. Biol. **18**, 69–88 (1983).
12. B.J. West, A.L. Goldberger, G. Rovner, V. Bhargava, "Nonlinear dynamics of the heartbeat. The AV junction: Passive conduict or active oscillator?", Physica D, **17**, 198–206 (1985).
13. H. Zhang, A.V. Holden, I. Kodama, H. Honjo, M. Lei, T. Varghese, M.R. Boyett, "Mathematical models of action potentials in the periphery and center of the rabbit sinoatrial node", Am. J. Physiol. Heart Circ. Physiol. **279**, H397–H421 (2000).
14. D. Postnov, Kee H. Seung, K. Hyungtae, "Synchronization of diffusively coupled oscillators near the homoclinic bifurcation", Phys. Rev. E **60**, 2799–2807 (1999).
15. B.F. Hoffman, P.F. Cranefield, *Electrophysiology of the Heart* (Mc Graw Hill, New York, 1960).
16. J. Jalife, V.A. Slenter, J.J. Salata, D.C. Michaels, Circ. Res. **52**, 642–656 (1983).
17. S.S. Demir, J.W. Clark, W.R. Giles, Am. J. Physiol. Heart. Circ. Physiol. **276**, H2221–H2244 (1999).

Part III
Cardiovascular Physics: Data Analysis

Part III
Cavity Chamber Physical Data Analysis

Statistical Physics of Human Heart Rate in Health and Disease

Ken Kiyono, Yoshiharu Yamamoto and Zbigniew R. Struzik

Abstract Complex phenomena know several benchmarks, or 'hard' and to date unsatisfactorily understood problems. Human heart rate control is such a complexity benchmark in biophysics, consistently defying full explanation. In our recent work, heart rate regulation by the autonomic nervous system has been shown to display remarkable fundamental properties of scale-invariance of extreme value statistics [1] in healthy heart rate fluctuations, ubiquitously observed in physical systems at criticality, and to undergo a phase transition as a result of altered neuro-regulatory balance [2]. Most recently, we have shown this behaviour to depart from the critical scale invariance in the case of life-threatening congestive heart failure (CHF). Our new index, derived from the non-Gaussianity characteristic, has proved to be the only meaningful one among all known heart rate variability (HRV) based mortality predictors [3]. We speculate on possible mechanisms for the increased variability and complexity of heart rate for life-threatening CHF, as reflected in the intermittent large deviations, forming non-Gaussian 'fat' tails in the probability density function (PDF) of heart rate increments and breaking the critical scale invariance observed in healthy heart rate [1, 2].

Keywords Congestive heart failure · Criticality · Heart rate variability · Intermittency · Non-Gaussianity

1 Introduction

Over the past thirty years, fluctuations in heart rate, often referred to as heart rate variability (HRV), have become a central topic in physiological signal analysis, serving as a vital non-invasive indicator of cardiovascular and autonomic system

K. Kiyono (✉)
College of Engineering, Nihon University, 1 Naka-gawara, Tokusada, Tamura-machi, Koriyama City, Fukushima, 963-8642, Japan
e-mail: kiyono@ge.ce.nihon-u.ac.jp

function. A Medline search reveals more than 8,000 papers to date on different aspects of HRV – this in a discipline which is subject to continued controversy and rapidly growing interest and scientific effort [4], closely following the exponential growth rate of the available computing power.

Methods employed mainly in statistical physics of complex phenomena and systems, including chaotic dynamics, have, in recent years, been increasingly used to analyse HRV time series. The demonstration of shared characteristics in HRV time series has encouraged comparisons across various phenomena, aiming at unravelling universality and suggesting the generic applicability of the underlying models. In particular, the multiplicative cascade picture, originally introduced as a model of hydrodynamic turbulence [5], has become an appealing paradigm to researchers to represent a generic underlying mechanism for a variety of complex phenomena. These have included the nature of autonomic regulation of human heart rate, where complexity characteristics such as long-range temporal anticorrelations [6–8], multifractal scaling properties [9, 10] and non-Gaussian probability density function (PDF) have been identified. Indeed, an explanation of heart rate complexity using the turbulence analogy has been proposed [11].

In order to attain a deeper insight into such parallels, and with the aim of understanding the multifractal and non-Gaussian nature of HRV time series, we consider a more detailed characterisation of intermittency and non-Gaussianity to be needed to understand the underlying mechanisms. To this end, we have introduced an analysis method [1, 12] to investigate deformation processes of the PDF of detrended increments when going from fine to coarse scales, to investigate non-Gaussian properties of the time series, which have not been systematically analysed in previous studies. Using this multiscale PDF analysis, we are in a position precisely to analyse scale dependence characteristics of the PDF of detrended time series. In particular, we are able to verify the multiplicative cascade hypothesis, or prove an alternative—the robust scale-invariance in the PDF of detrended time series characteristic of systems in the critical state.

The structure of the paper is as follows. In Section 2, we introduce multiscale PDF's and describe methods to quantify these PDF's. In Section 3, we analyse HRV in healthy subjects and compare artificially generated Gaussian and non-Gaussian noise. Then we demonstrate the robust scale-invariance of non-Gaussian PDF's of HRV, and discuss a criticality hypothesis of healthy human HRV in Section 4. In Section 5, we consider intermittency of HRV as a complexity measure, and show that increased complexity corresponds with high clinical risk, which contradicts the popularly suggested link between reduced complexity and clinical risk. Finally, in Section 6, we present our conclusions.

2 Multiscale PDF Analysis with Polynomial Detrending

To illustrate the analysis method of multiscale PDF's, we consider a noisy time series, $\{b(i)\}$ ($i = 1, \cdots, N_{\max}$). We first integrate the time series $\{b(i)\}$,

$$B(i) = \sum_{j=1}^{i} b(j), \quad (1)$$

and divide it into boxes of equal size s. In each interval $[1 + s(k-1), s(k+1)]$ of length $2s$, where k is the index of the box, we fit $B(i)$, using a polynomial function of the order d, which represents the trend in the corresponding segment. The differences $\Delta_s B(i) = B^*(i+s) - B^*(i)$ at a scale s are obtained in the corresponding subinterval $[1 + s(k-1), sk]$, where $B^*(i)$ is a deviation from the polynomial fit. By this procedure, the $(d-1)$-th order polynomial trends are eliminated. At the coarse-grained scale level s, the increment for each box is expressed as the local average:

$$\overline{\Delta_s B(i)} = \frac{1}{s} \sum_{j=1+s(k-1)}^{sk} \Delta_s B(i). \quad (2)$$

One of the widely used methods to characterise fractal signals with long-range power-law correlations and multiscaling (multifractal) properties of the amplitude is a scaling analysis of the q-th order partition function, $Z_q(s) = \langle |B(i+s) - B(i)|^q \rangle$, where $\langle \cdot \rangle$ denotes a statistical average [5, 13]. In the context of hydrodynamic turbulence, this function is referred to as the 'structure function'. For fractal signals, the structure function obeys the power law with s as $Z_q(s) \simeq s^{\zeta(q)}$. Further, for multifractal signals, these scaling exponents $\zeta(q)$ exhibit a non-linear dependence on q. Conventionally, the mean square displacement $Z_2(s)$ is related to the so-called Hurst exponent H as $Z_2(s) \sim s^{2H}$. The H represents the long-range power law correlation properties of the signal. If $H = 0.5$, there is no correlation and the signal $b(i)$, the increment of the analysed $B(i)$, is uncorrelated white noise; if $H < 0.5$, the signal is anti-correlated; if $H > 0.5$, the signal is correlated.

To establish how different the PDF is from a Gaussian distribution, we first obtain standardised PDF's (the variance has been set to one) of $\Delta_s B(i)$, and then use parameter estimation based on Castaing's equation introduced in the study of hydrodynamic turbulence [14].

To search for possible mechanisms behind complex fluctuations in HRV, a model to study hydrodynamic turbulence [14] is applied to the characterisation of the increment PDF's. We fit the increment PDF to the following function based on Castaing's equation:

$$\tilde{P}_s(x) = \int P_L\left(\frac{x}{\sigma}\right) G_{s,L}(\ln \sigma) d(\ln \sigma), \quad (3)$$

where P_L is the increment PDF at a large scale $L > s$, and the kernel $G_{s,L}$ (in log variables) determines the nature of the cascade-type multiplicative process. By the multiplication of x of (distribution) law $P_L(x)$, by σ of law $F_{s,L}(\sigma) = G_{s,L}(\ln \sigma)$, Castaing's equation (3) can be obtained. According to Castaing et al. [14], the cascade process defined by $G_{s,L}$ is self-similar if there is a decreasing sequence of scales $\{s_n\}$ such that

$$G_{s_0 s_n} = G_{s_0 s_1} \otimes G_{s_1 s_2} \otimes \cdots \otimes G_{s_{n-1} s_n}, \tag{4}$$

where $\otimes$ denotes the convolution product, and G_{s_{n-1},s_n} is independent of n. For a self-similar cascade process, the magnitudes of the increment are defined multiplicatively from coarse to fine scales in G kernel space. The notion of the multiplicative cascade processes has been considered as the paradigm of multifractal behaviour.

Here we assume P_L and $G_{s,L}$ are both Gaussian,

$$G_{s,L}(\ln \sigma) = \frac{1}{\sqrt{2\pi}\lambda} \exp\left(-\frac{\ln^2 \sigma}{2\lambda^2}\right), \tag{5}$$

and investigate the scale dependence of λ^2. The λ^2 is estimated from the minimum of the chi-square statistic χ^2 [15, 16]:

$$\chi^2 = \int \frac{(P_{\exp}(x) - P_{\text{fit}}(x))^2}{P_{\exp}(x) + P_{\text{fit}}(x)} dx, \tag{6}$$

where the $P_{\exp}$ and P_{fit} denote the observed and approximated PDF's, respectively.

Within the multiplicative cascade picture originally introduced as a model of hydrodynamic turbulence [5], the parameter λ^2 can be interpreted as being proportional to the number of cascade steps and is known to decrease linearly with $\log s$ for the self-similar cascade process [14, 17]. Naert et al. [15] demonstrate that the shape of the PDF is more sensitive to the variance of $G_{s,L}$ than to its exact shape. In the following, we demonstrate that the methodology of fitting Castaing's equation is useful in characterising intermittent and non-Gaussian fluctuations of a stochastic process different from the cascade processes.

3 Multiscale Characteristics of Healthy Human Heart Rate and Gaussian and non-Gaussian Noise

We analyse three sets of experimental data of HRV from 7 healthy subjects (7 males; ages 21–30 years) without any known disease affecting the autonomic control of heart rate. For the long-term heart rate data analysed here, the time series, $\{b(i)\}$, are sequential heart interbeat intervals $b(i)$, where i is the beat number. The first two sets of data consist of daytime (12:00–18:00 h, a length of up to 4×10^4 beats) and nighttime (24:00–06:00 h, a length of up to 3×10^4 beats) heart rate data, collected during normal daily life. The third set of experimental data consists of data of seven 26-hour long periods (up to 10^5 beats), collected when the subjects underwent constant routine (CR) protocol, where known behavioural factors affecting heart rate (e.g., exercise, diet, postural changes and sleep) are eliminated [10, 18].

In order to test for the possible presence of non-linear mechanisms in complex heart rate dynamics, we apply the surrogate data test to the CR protocol data [19]. We generate a surrogate set of data with the same Fourier amplitudes and distributions as the original increments in the CR protocol data. Since only linear temporal correlation of $b(i+1) - b(i)$ is retained in the surrogate data, a comparison with the raw data can be used to test whether the PDF of 'velocity' increments $\Delta_s B(i)$ possesses some non-linear mechanism inherent to it.

We also study the following examples of Gaussian and non-Gaussian noise: (1) an autoregressive model (AR model); (2) fractional Gaussian noise; (3) a multifractal heart rate model assuming a multiplicative cascade process (cascade model); (4) truncated Lévy noise (TL noise), and discuss the scale dependence of the increment PDF's for each process. For more details, the reader is referred to Ref. [12]. Representative examples of the increment series (2) are shown in Fig. 1. It is important to note that if we consider only the correlation properties of the fluctuations, we cannot see the significant difference in many models. For instance, the scaling exponents H_1 estimated using the mean square displacement $Z_2(s)$ are close to 1 for healthy human HRV and its models, such as the AR model and the cascade model, which means these models reproduce $1/f$ scaling in the power spectrum. However, an essential difference in these models is evident from the occurrence of large deviations. For the AR process driven by Gaussian noise (Fig. 1b) and fractional Gaussian noise (not shown), almost all increments stay in the range $[-3\sigma, 3\sigma]$, mainly in $[-2\sigma, 2\sigma]$, at all scales, because, for the Gaussian noise, the probability of the appearance of large increments exceeding 2σ is $\text{Prob}\{|\Delta_s B| > 2\sigma\} < 0.045$, $\text{Prob}\{|\Delta_s B| > 3\sigma\} < 0.0027$, and $\text{Prob}\{|\Delta_s B| > 4\sigma\} < 0.0001$. In comparison, for healthy HRV (e.g. Fig. 1a) we can see inhomogeneous and intermittent behaviour related to the high probability of the appearance of increments larger and smaller than the Gaussian noise. The heartbeat model based on the random cascade process (Fig. 1c) and the truncated Lévy noise (Fig. 1d) also show such intermittent behaviour.

The intermittent behaviour is related to the non-Gaussian PDF (Fig. 1g,h) with marked fat tails and a peak around the mean value, which indicates a high probability of the appearance of larger and smaller increments than the Gaussian noise. As shown in Fig. 1e, the PDF of the HRV is also non-Gaussian with stretch-exponential-like fat tails, which is different from the quadratic function of the logarithm of the Gaussian PDF as expected for the AR process (Fig. 1f).

Although the cascade heartbeat model and the uncorrelated noise with a truncated Lévy distribution follow different mechanisms, the non-Gaussian PDF's of both the processes (Fig. 1g,h) at certain scales are similar to that of the HRV (Fig. 1e). This means that we cannot fully characterise the dynamics of HRV only from the shape of the PDF. However, in terms of the scale dependence of the multiscale PDF, HRV shows an important difference compared with the other non-Gaussian models. In general, we can expect Gaussian PDF's at sufficiently large scales because of the accumulation of uncorrelated variations (as described by the central limit theorem). Thus, the non-Gaussian process at a fine scale progressively

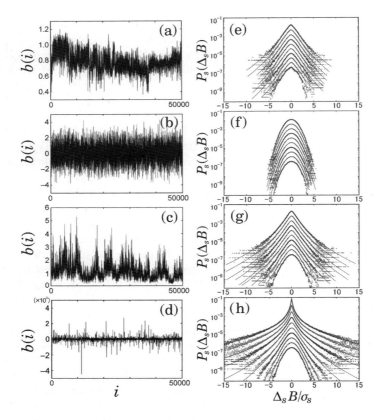

Fig. 1 (*left*) (**a**) A representative record of daytime (12:00–18:00 h) heart interbeat intervals of a healthy subject; (**b**) an autoregressive model (order 10); (**c**) a multifractal heart rate model assuming a multiplicative cascade process; (**d**) independent, identically distributed random variables with a truncated Lévy distribution for $\gamma = 0.001$, and $\alpha = 0.75$. (*right*) Deformation of increment PDF's across scales. Standardised PDF's (in logarithmic scale) of $\Delta_s B(i)$ for different time scales are shown for (from *top* to *bottom*) $s = 8, 16, 32, 64, 128, 256, 512, 1024$ beats: (**e**) daytime (12:00–18:00 h) heart interbeat intervals for healthy subjects (7 males; ages 21–30 years); (**f**) an autoregressive model (order 10); (**g**) a multifractal heart rate model assuming a multiplicative cascade process; (**h**) independent, identically distributed random variables with a truncated Lévy distribution. These PDF's are estimated from all samples for each group. In solid lines, we superimpose the deformation of the PDF using Castaing's equation

converges to the Gaussian, as the scale s increases. In practice, we can observe this convergence process for the cascade model and the truncated Lévy noise (Fig. 1g,h).

In contrast, the PDF's of HRV at different scales retain an invariant shape [1], as revealed by the scale invariance of λ^2 (Fig. 2a). This feature is more clearly seen in the HRV during constant routine protocol [12]. Moreover, the non-Gaussian

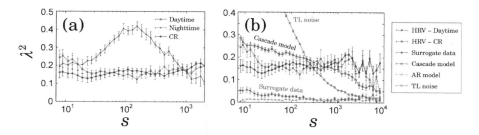

Fig. 2 Scale-dependence of a fitting parameter of Castaing's equation, λ^2. (**a**) The averaged values over 7 healthy subjects; comparison between daytime during normal daily life, with constant routine protocol, and night-time (24:00–06:00 h). (**b**) Comparison with artificially generated Gaussian and non-Gaussian noise (see text for details). The error bars indicate the standard error of the group averages

PDF's for the daytime and constant routine show a striking resemblance. This scale invariance in the non-Gaussian PDF disappears for the night-time HRV [12]. This fact suggests that the difference in the neuroautonomic control mechanisms between daytime and night-time may be the cause of the difference in the scale-invariant PDF. For the night-time HRV, Non-Gaussian fluctuations at a characteristic scale $\simeq 100$ beats are dominant.

Strictly speaking, the function of the PDF for the truncated Lévy distribution is not exactly the same form of the approximated PDF based on Castaing's equation. However, through the convergence process to the Gaussian, as discussed in Ref. [12], the χ^2 – the difference between the truncated Lévy distribution and the approximated PDF - becomes much smaller according to a power law decay. In our analysis of the long-term HRV, the χ^2 corresponding with the optimal fit of the approximated PDF is in the order of 10^{-4}–10^{-5}. This means that the very small values of the χ^2 ($< 10^4$) make it impossible to distinguish between the truncated Lévy distribution and the approximated PDF based on Castaing's equation. Thus, the convergence process of the truncated Lévy noise is also well characterised by a single parameter λ^2.

The main difference between the cascade-type model, including higher order correlations, and uncorrelated noise with the truncated Lévy distribution is the scale-dependence of the parameter λ^2. For the cascade-type model, $\lambda^2 \simeq \log s$, as generally expected for random cascade models, and for the truncated Lévy noise, a power law decay (Fig. 2b) is obtained, as expected from the Berry-Esséen theorem [20], which states that the rate of convergence to the Gaussian is in the order of $s^{-1/2}$.

Surprisingly, the multiscale PDF's of HRV during the daytime do not show a convergence to the Gaussian over a wide range of scales [1], although the PDF's of the surrogate data are near Gaussian and progressively converge to the Gaussian (Fig. 2b). The range of scales where this scale-invariance of non-Gaussian PDF is observed, spanning from about 10 beats to a few thousand heartbeats, is compatible with that of the robust, behavioural-independent $1/f$ scaling [18] and multifractality

[10] of heart rate. The scale invariance in the PDF is also robust in the sense that it is observed not only during CR protocol but also during normal daily life, where behavioural modifiers of heart rate dramatically change the mean level of heart rate. We have thus established a novel property of scale invariance in healthy human heart rate dynamics [1].

It has recently been suggested that the underlying mechanism of HRV fluctuations shares the general principles of other critical point phenomena in complex systems [11], because long-range correlation ($1/f$ fluctuation) and multifractality are typical characteristics of a system at the critical point [21, 22]. In addition, we have found the scale-invariant properties in the non-Gaussian PDF to be one characteristic feature observed at the critical point, and to support the criticality hypothesis [1]. The critical behaviour occurs at the transition point between two different phases, and it has recently been observed in a variety of transport and network dynamics, such as vehicular traffic flow [23], Internet traffic [24] and granular flows [25].

In such transport systems, the critical point is the phase transition point from an 'uncrowded' state to a 'congested' state in the transport routes. A functional advantage of the system being at or near the critical point is that maximum efficiency of transport is realised [24]. In the case of the heart, these states correspond with a state with an insufficient pre-load due to central hypovolaemia (uncrowded) and a state with an excessive after-load due to 'congestive' heart failure (congested). The two cases are known to be associated with decreased cardiac output due to a lack of blood to pump out and difficulty in pumping out the blood, respectively [26]. Therefore, our results may indicate that the central neuroregulation continually brings the heart to a critical state to maximise its functional ability. This may be particularly important in understanding the widely reported evidence that the decreased $1/f$ scaling in the low frequency region of the power spectrum [27, 28] is associated with increased mortality in cardiac patients. Our findings suggest that a breakdown of the optimal control achieved in the critical state may be related to this clinically important phenomenon. Further, in Kiyono et al. [2], we experimentally confirm the existence of a second order phase transition in healthy heart rate.

4 Phase Transition in Healthy Human Heart Rate

As previously reported in Kiyono et al. [2], we analyse seven records of healthy subjects (mean age: 25.3 yr) in three behavioural states: (1) usual daily activity; (2) experimental exercise; and (3) sleep. The data set consists of the interbeat intervals between consecutive heartbeats measured over 24 h (Fig. 3), in which the subjects were initially asked to ride on a bicycle ergometer for 2.5 hours, as the exercise state, and maintain their heartbeat at $500 \simeq 600$ ms. After the exercise, the data were continuously measured during usual daily activity in the daytime and sleep at night, with regular sleep schedules. As shown in Fig. 3, we analyse the data from four intervals: (A) constant exercise; (B) usual daily activity after the exercise; (C) sleep;

Fig. 3 A representative record of heart interbeat intervals for a healthy subject measured over 24 h (13:00). For analyses, we selected four subintervals during different states from the records of seven healthy subjects

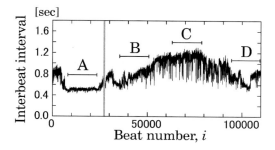

and (D) usual daily activity the next morning. The data in each interval contains over 15,000 heartbeats.[1]

In order to characterise the heart rate fluctuations, we use a detrended random walk method, as outlined in Section 2. For a quantitative comparison, we fit the data to the function (3) and estimate the variance λ_s^2 of ω_s in (3) [14]. The deformation of the standardised PDF's of the increment $\Delta_s B$ across scales is shown in Fig. 4 for all four intervals A,B,C,D from Fig. 3, showing the non-Gaussian PDF, which is well described by Eq. (3), as reported in our previous study [1]. The scale dependence of λ^2, as shown in Fig. 5, indeed exhibits distinctive features [2]. The non-Gaussian PDF's for the two records of daily activity show a striking resemblance. The scale invariance of the non-Gaussian PDF in a range of $20 - 1,000$ beats, however, disappears in sleep states, in which non-Gaussian fluctuations at a characteristic scale of $\simeq 100$ beats are dominant. The λ^2 for constant exercise is much smaller than for the other states, implying near Gaussianity.

We thus demonstrate that healthy human heart rate exhibits phase transition-like dynamics between different behavioural states, with a dramatic departure from criticality. Strongly correlated fluctuations—a hallmark of criticality—are observed only in the narrow region of usual daily activity, while a breakdown follows the transition to other states. This transition is reminiscent of a second order phase transition, and supports the hypothesis [1] that healthy human heart rate is controlled to converge continually to a critical state during usual daily activity. We believe this feature to be a key to understanding the heart rate control system [30].

5 Increased non-Gaussian Temporal Heterogeneity of Heart Rate in Fatal Heart Failure

Inspired by the intermittency problem in fully developed turbulence [14], we investigate the PDF of multiscale HRV [1, 31] of patients with congestive heart failure (CHF), and show *increased* intermittency characterised by strongly non-Gaussian

---

[1] The methods of data collection and pre-processing are the same as those in Ref. [29]. All the subjects gave their informed consent to participating in this institutionally approved study. The data is available at http://www.p.u-tokyo.ac.jp/k_kiyono

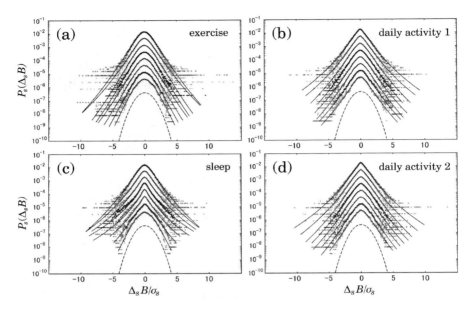

Fig. 4 Deformation of standardised PDF's (the standard deviation has been set to one) of the increment $\Delta_s B$ across scales, where the σ_s is the standard deviation of $\Delta_s B(i)$. The standardised PDF's (in logarithmic scale) for different time scales are shown for (from top to bottom) $s = 8, 16, 32, 64, 128, 256, 512, 1024$ beats. The PDF's are estimated from all the records for each condition: (**a**) constant exercise; (**b**) usual daily activity after the exercise; (**c**) sleep; (**d**) usual daily activity the next morning. The dashed line is a Gaussian PDF for comparison. In solid lines, we superimpose the deformation of the PDF using Castaing's equation [14] with the log-normal self-similarity kernel, providing an excellent fit to the data

PDF predicts the mortality of patients with CHF more strongly than previously introduced HRV measures [3].

We evaluate a 24-hour Holter electrocardiogram of 108 CHF patients, of whom 39 patients (36.1%) died within four years of the follow-up period.[2] During the follow-up, patients or their families were periodically sent questionnaires or interviewed by telephone. The end point was all-cause death, but the majority (35/39) was cardiac death, including death from progressive heart failure, sudden death and acute myocardial infarction.

From January 2000 to December 2001, we prospectively enrolled patients who were consecutively admitted to hospital for worsening heart failure. Twenty-four-hour monitoring of Holter electrocardiogram (ECG) was conducted before the patients received vasoactive intravenous therapy. Most of the patients were resting in bed during the Holter recording. During the recruitment period, 215 patients were

---

[2] All the patients gave their informed, written consent. The study protocol was approved by the ethics committee of the Fujita Health University Hospital.

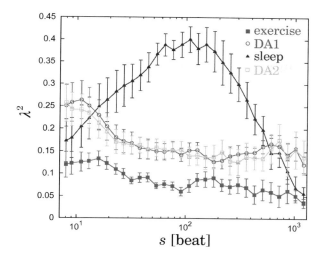

Fig. 5 Dependence of the fitting parameter λ^2 on the scale s during constant exercise, usual daily activity after the exercise (DA1), sleep, and usual daily activity the next morning (DA2)

assessed for enrolment eligibility. One hundred and seven patients were excluded due to predetermined criteria: 24 h Holter ECG not being analysed because noise or non-sinus beats represented > 5% of the total monitoring time ($n = 7$), chronic or paroxysmal atrial fibrillation ($n = 80$), a permanent or temporary cardiac pacemaker ($n = 18$), active thyroid disease ($n = 1$), or malignancy ($n = 1$).

Two-channel (NASA and CM5) Holter ECG was recorded 1 ± 1 day after admission to hospital. The ECG signals were digitised at 125 Hz and 12 bits, and processed off-line using a personal computer equipped with dedicated software (Cardy Analyzer II, Suzuken, Japan). Only recordings with at least 22 hours of data and > 95% of quantified sinus beats were included in the analysis. All the QRS complexes in each recording were detected and labelled automatically. The results of the automatic analysis were reviewed, and any errors in R wave detection and QRS labelling were edited manually. Then computer files were generated containing the duration of individual R-R intervals and classifications of individual QRS complexes (normal, supra-ventricular and ventricular premature complexes, supraventricular and ventricular escape beats). The frequency of ventricular premature beats (VPB's) per hour was also calculated.

The 24 h sequence of the intervals between two successive R waves of sinus rhythm, i.e. HRV, was analysed. To avoid the adverse effects of any remaining errors in the detection of the R wave, large (> 20%) consecutive R-R interval differences were thoroughly reviewed until all errors were corrected. In addition, when atrial or ventricular premature complexes were encountered, the corresponding R-R intervals were interpolated by the median of the two successive beat-to-beat intervals.

In Fig. 6(a,b), representative examples of HRV dynamics for a survivor and a non-survivor are shown. The non-survivor has higher, *not lower*, R-R intervals and standard deviation of normal inter-beat intervals (SDNN) compared with the survivor. Also, temporal heterogeneity of local variance is very significant in the non-survivor, suggesting increased intermittency of the signal. Such intermittency can also be seen in the time series of $\Delta_{40}B(i)$ (Fig. 6c,d). In the time series of $\Delta_{1000}B(i)$, the temporal heterogeneity of local variance in the non-survivor is stronger than that in the survivor (Fig. 6e,f). The intermittent large deviations of HRV form non-Gaussian fat tails and a peak around the mean value in the PDF's of $\Delta_s B(i)$, Fig. 6(g,h), particularly that of $\Delta_{40}B(i)$. When the λ_s is compared between groups of survivors and non-survivors at each scale s (Fig. 7a), the average λ_s values are consistently higher in the non-survivors than in the survivors, implying higher intermittency of HRV in the non-survivors. Significant differences are observed in the range of [15, 200] beats and [400, 1200] beats.

We thus find that CHF patients with intermittent HRV have a higher risk of mortality compared with patients with less intermittency. It is evident from Fig. 7a that the multiscale non-Gaussianity index λ, and in particular the short-term (< 40 beats) non-Gaussianity λ_{40}, probing the existence of intermittent, episodically large fluctuations of heart rate, is a significant risk stratifier for CHF mortality. It is further evident that CHF patients at high mortality risk have a heart rate of higher intermittent variability than CHF survivors (and healthy people). This is in stark contrast to the emergent picture entailing a lower complexity paradigm associated with CHF, repeatedly suggested in numerous studies. This is well illustrated by the behaviour of the λ_s for CHF patients' data from Physionet, widely used in physics literature as a benchmark for complexity of heart rate and related measures and analysis methodologies (Fig. 7b). Therefore, we consider that the reduced heart rate variability/complexity paradigm may lead to a dangerous true-negative result in CHF risk stratification and should thus be reappraised.

As outlined in previous sections, we have found a non-Gaussian property in the multiscale PDF of HRV and its robust scale-invariance (scale-independence) in healthy individuals [1,2]—a hallmark of complex dynamics at criticality. Compared with healthy human HRV, heart rate fluctuations in CHF patients, especially those in non-survivors, are characterised by a scale specific increase of non-Gaussianity in the short scale (< 40 beats) (Fig. 7a). One possible implication of the increased short scale non-Gaussianity in CHF patients, particularly non-survivors, is a selective breakdown in the short-term neural regulation of heart rate (i.e. baroreflexes); insufficiency or instability in the negative feedback control may lead to HRV dynamics with unresponsive quiescent phases—giving rise to cardiac congestion—interwoven with intermittent compensatory bursts. The fact that CHF is characterised by impaired or down-regulated baroreflex systems [32] also supports this view. Although the physiological origin of the short-term non-Gaussianity is still undetermined, there are many physical, as well as non-physical, systems showing this type of intermittent strong non-Gaussianity with temporal heterogeneity of local variance. Drawing analogies with such systems might help in understanding this. In the study of hydrodynamic turbulence, the non-Gaussian

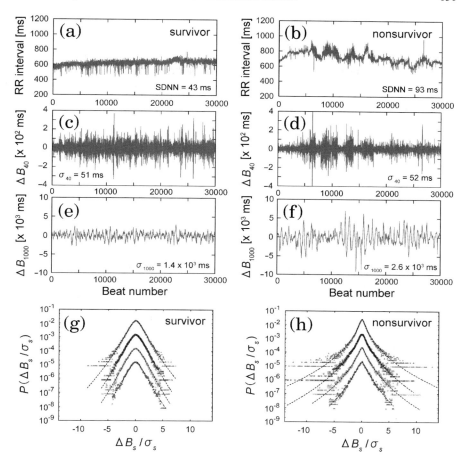

Fig. 6 Multiresolution PDF characterisation of R-R interval fluctuations. Illustrative examples of time series of R-R intervals (**a, b**), time series of $\Delta_{40}B(i)$ (**c, d**), time series of $\Delta_{1000}B(i)$ (**e, f**), and standardised PDF's (in logarithmic scale) of $\Delta_s B(i)$ for (from the *top* to *bottom*) $s = 40, 100, 300, 1000$ beats (**g, h**), where σ_s denotes the standard deviation of $\Delta_s B(i)$. In dashed lines, we superimpose the PDF approximated by Castaing's model [14]. The panels on the left (**a, c, e, g**) are data for a male survivor during a follow-up period of 43 months. The panels on the right (**b, d, f, h**) are data for a female non-survivor who died suddenly 2 months after the measurement

PDF of the intermittent fluctuations has been demonstrated to result from clustering of the local variance as reflected in magnitude correlations [33]. In Kiyono et al. [2], we have elucidated an underlying cause of such intermittent clustering of variance in HRV dynamics by analysing magnitude correlations in healthy individuals.

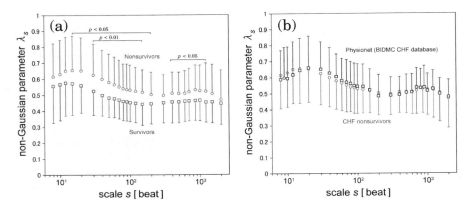

Fig. 7 Scale dependence of non-Gaussian parameter λ_s. . (**a**) Significant differences between the survivors and non-survivors are observed in the range of [15, 200] beats and [400, 1200] beats. The squares represent the averages for 69 survivors and the circles those for 39 non-survivors. The error bars are the standard deviation. (**b**) The λ_s for Physionet data base of CHF individuals compared with non-survivors from our study

6 Conclusions

We describe the methodology of multiscale PDF analysis of time series from complex phenomena suitable for assessing the nature of the underlying process.

For the example of the multiscale PDF properties of healthy heart rate regulation, we suggest, as previously stated in Kiyono et al. [1, 2], that the theory of phase transitions and critical phenomena in non-equilibrium systems [23, 34] may be useful in elucidating the mechanism of complex heart rate dynamics. In particular, we show that, contrary to the multiplicative cascade hypothesis, the multiscale PDF's of the HRV during the daytime do not show a convergence to the Gaussian over a wide range of scales. Surprisingly, the scale-independent and fractal structure is preserved not only in a quiescent condition, but also in a dynamic state of daily life, in which the mean level of heart rate is dramatically changing [1]. This invariance property suggests that the underlying mechanism of HRV fluctuations shares the general principles of other critical point phenomena in complex systems, including the possibility of phase transition [2].

A breakdown of the optimal control achieved in the critical state, which we observe in increased complexity corresponding with increased short-term intermittency and non-Gaussianity in high risk CHF patients, is a clinically important finding. The non-Gaussianity index λ_s appears to detect controller instability and thus is thought significantly to reflect increased risk of death in patients with CHF.

In summary, the new index, λ_s, detects fundamental characteristics of heart rate fluctuations that are not well captured by conventional HRV analysis.

References

1. Kiyono, K., Struzik, Z.R., Aoyagi, N., Sakata, S., Hayano, J., Yamamoto, Y.: Critical scale-invariance in healthy human heart rate. Phys. Rev. Lett. **93** (2004) 178103
2. Kiyono, K., Struzik, Z.R., Aoyagi, N., Togo, F., Yamamoto, Y.: Phase transition in healthy human heart rate. Phys. Rev. Lett. **95** (2005) 058101
3. Kiyono, K., Hayano, J., Watanabe, E., Struzik, Z.R., Kodama, I., Hishida, H., Yamamoto, Y.: Non-Gaussian heart rate as an independent predictor of mortality of chronic heart failure patients. submitted (2007)
4. Parati, G., Mancia, G.: Point:counterpoint: Cardiovascular variability is/is not an index of autonomic control of circulation. J. Appl. Physiol. **101** (2006) 676–682
5. Frisch, U.: Turbulence. 1 edn. Cambridge University Press, Cambridge (1995)
6. Peng, C.K., Mietus, J., Hausdorff, J.M., Havlin, S., Stanley, H.E., Goldberger, A.L.: Long-range anticorrelations and non–Gaussian behavior of the heartbeat. Phys. Rev. Lett. **70** (1993) 1343–1346
7. Yamamoto, Y., Hughson, R.L.: On the fractal nature of heart rate variability in humans: effects of data length and β-adrenergic blockade. Am. J. Physiol. **266** (*Regulatory Integrative Comp. Physiol.* 35) (1994) R40–R49
8. Peng, C.K., Havlin, S., Stanley, H.E., Goldberger, A.L.: Quantification of scaling exponents and crossover phenomena in nonstationary heartbeat time series. Chaos **5** (1995) 82–87
9. Ivanov, P.C., Amaral, L.A.N., Goldberger, A.L., Havlin, S., Rosenblum, M.G., Struzik, Z.R., Stanley, H.E.: Multifractality in human heart rate dynamics. Nature **399** (1999) 461–465
10. Amaral, L.A.N., Ivanov, P.C., Aoyagi, N., Hidaka, I., Tomono, S., Goldberger, A.L., Stanley, H.E., Yamamoto, Y.: Behavioral-independent features of complex heartbeat dynamics. Phys. Rev. Lett. **86** (2001) 6026–6029
11. Lin, D.C., Hughson, R.L.: Modeling heart rate variability in healthy humans: A turbulence analogy. Phys. Rev. Lett. **86** (2001) 1650–1653
12. Kiyono, K., Struzik, Z.R., Aoyagi, N., Yamamoto, Y.: Multiscale probability density function analysis: Non-Gaussian and scale-invariant fluctuations of healthy human heart rate. IEEE Trans. Biomed. Eng. **53** (2006) 95–102
13. Barabási, A.L., Vicsek, T.: Multifractality of self-affine fractals. Phys. Rev. A **44** (1991) 2730–2733
14. Castaing, B., Gagne, Y., Hopfinger, E.J.: Velocity probability density-functions of high Reynolds-number turbulence. Physica D **46** (1990) 177–200
15. Naert, A., Puech, L., Chabaud, B., Peinke, J., Castaing, B., Hebral, B.: Velocity intermittency in turbulence: how to objectively characterize it? J. Phys. II France **4** (1994) 215–224
16. Press, W.H., Teukolsky, S.A., Vetterling, W.T., Flannery, B.P.: Numerical Recipes in C. 2 edn. Cambridge University Press, Cambridge (1992)
17. Ghashghaie, S., Breymann, W., Peinke, J., Talkner, P., Dodge, Y.: Turbulent cascades in foreign exchange markets. Nature **381** (1996) 767–770
18. Aoyagi, N., Ohashi, K., Yamamoto, Y.: Frequency characteristics of long-term heart rate variability during constant routine protocol. Am. J. Physiol. **285** (2003) R171–R176
19. Schreiber, T., Schmitz, A.: Surrogate time series. Physica D **142** (2000) 346–382
20. Feller, W.: An Introduction to Probability Theory and Its Applications. 2 edn. Volume 2. Wiley, New York (1971)
21. Hinrichsen, H., Stenull, O., Janssen, H.K.: Multifractal current distribution in random-diode networks. Phys. Rev. E **65** (2002) 045104
22. Kadanoff, L.P., Nagel, S.R., Wu, L., Zhou, S.: Scaling and universality in avalanches. Phys. Rev. A **39** (1989) 6524–6537
23. Chowdhury, D., Santen, L., Schadschneider, A.: Statistical physics of vehicular traffic and some related systems. Phys. Rep. **329** (2000) 199–329
24. Takayasu, M., Takayasu, H., Fukuda, K.: Dynamic phase transition observed in the Internet traffic flow. Physica A **277** (2000) 248–255

25. Hou, M., Chen, W., Zhang, T., Lu, K.: Global nature of dilute-to-dense transition of granular flows in a 2D channel. Phys. Rev. Lett. **91** (2003) 204301
26. Guyton, A.C., Hall, J.E.: Textbook of Medical Physiology. 10 edn. Elsevier Science, Amsterdam (2000)
27. Bigger, Jr., J.T., Steinman, R.C., Rolnitzky, L.M., Fleiss, J.L., Albrecht, P., Cohen, R.J.: Power law behavior of RR-interval variability in healthy middle-aged persons, patients with recent acute myocardial infarction, and patients with heart transplants. Circulation **93** (1996) 2142–2151
28. Bigger Jr., J.T., Fleiss, J.L., Steinman, R.C., Rolnitzky, L.M., Kleiger, R.E., Rottman, J.N.: Frequency domain measures of heart period variability and mortality after myocardial infarction. Circulation **85** (1992) 164–171
29. Aoyagi, N., Ohashi, K., Tomono, S., Yamamoto, Y.: Temporal contribution of body movement to very long-term heart rate variability in humans. Am. J. Physiol. Heart Circ. Physiol. **278** (2000) H1035–H1041
30. Struzik, Z.R., Hayano, J., Sakata, S., Kwak, S., Yamamoto, Y.: $1/f$ Scaling in heart rate requires antagonistic autonomic control. Phys. Rev. E **70** (2004) 050901(R)
31. Kiyono, K., Struzik, Z.R., Yamamoto, Y.: Criticality and phase transition in stock-price fluctuations. Phys. Rev. Lett. **96** (2006) 068701-1–4
32. Packer, M.: Neurohormonal interactions and adaptations in congestive heart failure. Circulation **77** (1988) 721–730
33. Arneodo, A., Bacry, E., Manneville, S., Muzy, J.F.: Analysis of random cascades using space-scale correlation functions. Phys. Rev. Lett. **80** (1998) 708–711
34. Sethna, J.P., Dahmen, K.A., Myers, C.R.: Crackling noise. Nature **410** (2001) 242–250

Cardiovascular Dynamics Following Open Heart Surgery: Early Impairment and Potential for Recovery

Robert Bauernschmitt, Niels Wessel, Hagen Malberg, Gernot Brockmann, Jürgen Kurths, Georg Bretthauer and Rüdiger Lange

Abstract Heart rate variability (HRV) and baroreflex sensitivity (BRS) are severely altered during the postoperative course of patients undergoing cardiovascular surgery. This study was performed to evaluate the response of the autonomic regulation in patients undergoing cardiac surgery with a special emphasis on the potential for recovery.

Keywords Heart rate variability · Baroreflex sensitivity · Cardiac surgery · Postoperative autonomic regulation

1 Introduction

It is well known from earlier studies that the state of the autonomous tone has a major impact on survival and the occurrence of arrhythmias in patients after myocardial infarction [1, 2] and is severely altered in patients with dilated cardiomyopathy [3]. Analysing the response to open heart surgery with heart-lung machine, it was demonstrated that overall heart rate variability (HRV) is reduced after coronary artery bypass graft (CABG) surgery and has a tendency to recover within 3–6 months [4]. Recent studies proposing a more comprehensive evaluation of the autonomic system including evaluation of baroreflex sensitivity and nonlinear dynamics can increase the information obtained by traditional time and frequency domain HRV measures [5]. In this study we used HRV, nonlinear dynamics and spontaneous baroreflex parameters for evaluating the postoperative course of the autonomic response of unselected patients undergoing cardiovascular surgery. Aim of the study was to describe a typical pattern of postoperative changes to increase knowledge as a basis for further studies tending towards individual risk stratification.

R. Bauernschmitt (✉)
Department of Cardiovascular Surgery, German Heart Centre, Lazarettstraße 36, 80636 Munich, Germany
e-mail: bauernschmitt@dhm.mhn.de

2 Methods

2.1 Patients and Operations

207 consecutive patients undergoing various types of open heart surgery were included. Patients with documented atrial fibrillation, history of more than 10% ectopic beats, implanted pacemakers and recording times shorter than 10 minutes were excluded from this study, so 166 patients remained for analysis. 99 men and 67 women were included; their age being between 25 and to 85 years. 93 were undergoing valve replacement and 46 coronary artery bypass (CABG) surgery. 27 patients had combined procedures. All patients underwent conventional extracorporeal circulation (ECC) in mild hypothermia; the duration of extracorporal circulation ranged between 44 and 204 minutes. Anaesthesia was induced with midazolam, sufentanil and pancuronium and was maintained with isoflurane and with continuous infusion of fentanyl and midazolam. Moderate systemic hypothermia (core temperature 30–33°C) was used. Cardioplegic arrest was induced by antegrade cold crystalloid cardioplegia or blood cardioplegia during cardiopulmonary bypass.

After surgery, all patients were mechanically ventilated until achieving stable hemodynamics and the ability of spontaneous breathing. Piritramid was administered intravenously in 5-mg increments for pain control. Preoperative medication was continued as soon as possible.

3 Measurements

After an equilibration period of 10 minutes patients lay in supine position with both arms parallel to the body throughout the measurements. At the beginning of the measurements patients were connected to the following equipment: The Colin CBM-7000 system manufactured by the Colin Corporation in Japan. The Colin tonometer uses a pressure transducer (linear array of 15 pressure sensitive piezoelectric elements) applied firmly to the skin surface pressing and partially flattening an artery on to a bone surface. Arterial pressure waveforms are transmitted directly to the transducer. The system was calibrated by systolic and diastolic measurements from a standard arm cuff by oscillometric technique. As default, the instrument performs a calibration at regular intervals (2.5 and 5 min). Continuous non invasive blood pressure and ECG were sampled during a 30 min recording period. Special care was taken to perform the measurements (1d pre-OP, 1d post-OP, 7d post-OP and 3 months post-OP) during the same time of the day in each patient. The signals were collected at 1000 Hz and channelled into a computer (bed side laptop) by an analogue-to-digital signal converter. For data preprocessing, original data were filtered removing premature beats, artefacts and noise, keeping the original time reference. The RR-intervals recognized as not normal were treated as follows: removal from the series, linear interpolation or spline interpolation. The disadvantage of simply removing the beats is the loss of time dependence which can cause estimates of artificial frequencies. Interpolating linearly may lead to false decreased variability, interpolating with splines may fail in time series including a lot of ventricular premature beats (VPCs). As proved in several clinical studies, an adaptive filtering algorithm was able to exclude premature beats and artifacts [3]. The main advantage of this procedure is the spontaneous adaptation to variability changes in the series, which enables a more reliable removal of artifacts and VPCs. This new filtering algorithm consists of three sub-procedures: (i) the removal of obvious recognition errors, (ii) the adaptive percent filter, and (iii) the adaptive controlling filter.

3.1 Baroreflex Sensitivity (BRS): Dual Sequence Method (DSM)

The BRS is defined as the change of the heart frequency (beat-to-beat interval) related to increasing or decreasing values in systolic blood pressure and is expressed in ms/mmHg. The most relevant parameters for estimating the spontaneous baroreflex are the slopes of the regression line between SBP (systolic blood pressure) and BBI (beat-to-beat-interval). The DSM is based on standard sequence methods with several modifications. Two types of beat to beat interval responses were analyzed:

bradycardic: blood pressure increase causes RR-interval increase,
tachycardic: blood pressure decrease causes RR-interval decrease.

The bradycardiac fluctuations represent the vagal spontaneous regulation. The delayed tachycardic responses of heart rate (shift 3) are assigned to the slower beginning of sympathic regulation. The following parameter groups are calculated by DSM: (1) the total numbers of slopes in different sectors within 30 min; (2) the percentage of the slopes in relation to the total number of slopes in the different sectors; (3) the numbers of bradycardic and tachycardic slopes; (4) the shift operation from the first (sync mode) to the third (shift 3 mode) heartbeat triple; and (5) the average slopes of all fluctuations (see Fig. 1 for further clarification). DSM parameters are defined as described by Malberg et al. [6].

3.2 Heart Rate Variability (HRV)

HRV was assessed by time- and frequency-domain analyses. In time domain analysis, the intervals between adjacent normal R waves (NN intervals) are measured over the period of recording. A variety of statistical variables can be calculated from the intervals directly and others can be derived from the differences between

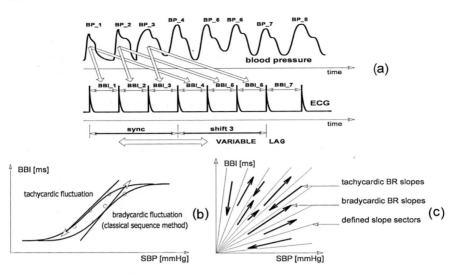

Fig. 1 Baroreflex calculation using the Dual Sequence Method (DSM). (**a**) Changes in beat-to-beat interval (BBI) in response to spontaneous fluctuations in blood pressure are calculated, either synchronous (sync) or in a delayed mode (shift 3). (**b**) Increases of systolic blood pressure cause increases in BBI (parasympathetically mediated), while drops of pressure result in decreased BBI (sympathetically mediated). Both components can be monitored by the DSM. (**c**) the distribution of slope sectors can be quantified

the intervals. Since many of the measures correlate closely with others, the suggestions by the Task Force HRV were respected and the following standard parameters calculated [7]:

> meanNN (mean value of normal beat-to-beat intervals): is inversely related to mean heart rate,
> sdNN (standard deviation of intervals between two normal R-peaks): gives an impression of the overall circulatory variability,
> rmssd (root mean square of successive RR-intervals): higher values indicate higher vagal activity.

Spectral analysis converted information in the time domain into information in the frequency domain. The most widely used method for processing the studied signal is the Fast Fourier Transformation. The HRV analysis focuses especially on two frequency bands of the power spectrum, high-frequency components (HF, 0.15–0.4 Hz, high values indicate vagal activity) and low-frequency components (LF, 0.04–0.15 Hz, high values indicate sympathetic activity). The following ratios were considered:

> LFn – the normalized low frequency (LFn = LF/(LF + HF)),
> HP/P – the to the total power P normalized high frequency as well as
> LP/P – the P normalized low frequency.

3.3 Nonlinear Dynamics

Heart rate and blood pressure variability and the baroreflex response represent complex interactions of many different control loops of the cardiovascular system. For control of the sinus node activity modulation system, predominantly non-linear behaviour is to be assumed. Thus, even the detailed description and classification of dynamic changes by time and frequency domain parameters may not be sufficient.

Complex processes, interrelations and new parameters can be described by methods of nonlinear dynamics. Condition is the transformation of the time series into symbol sequences with predefined symbols. In this process some detailed information is lost but the coarse dynamic behaviour can be analysed. Several new measures of non-linear dynamics in order to distinguish different types of heart rate dynamics as proposed by Kurths [8] and Voss [9] were used.

The concept of symbolic dynamics is based on a coarse-graining of dynamics. The difference between the current value (BBI or systolic blood pressure) and the mean value of the whole series is transformed into four symbols (0; 1; 2; 3). Symbols '0' and '2' reflect low deviation (decrease or increase) from mean value, whereas '1' and '3' reflect a stronger deviation (decrease or increase over a predefined limit); for details see Voss et al. [9]. Further, the symbol string can be transformed to 'words' of three successive symbols explaining the nonlinear properties and thus the complexity of the system. A high percentage of words consisting only of the symbols '0' and '2' ('wpsum02') reflects decreased HRV.

4 Statistics

All statistical analyses were performed by software SPSS 11.5 (SPSS Inc., Chicago, IL, U.S.A.). Mann-Whitney-U-test was applied to find differences within the calculated parameters. Results are given as mean $+/-$ standard deviation.

5 Results

Figures 2 and 3 exemplarily show the tracings of systolic and diastolic pressure (first panel) as well as of the RR-intervals (second). Figure 2 represents the pre-operative measurements, while the curves in Figure 3 have been recorded post-operatively. Pure visual inspection gives the impression of an overall reduced variability after the surgical procedure. As described before, these tachograms are the basis for subsequent analyses.

Time domain parameters of heart rate variability showed (Fig. 4) a significant decline 24 h after surgery, a visible tendency to restoration after one week and a significant recovery to more or less normal levels three months after surgery.

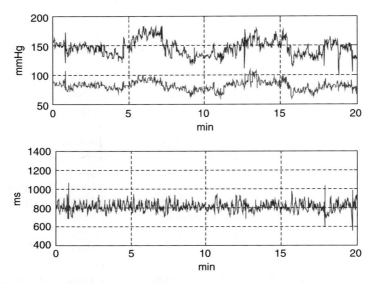

Fig. 2 Tachograms of blood pressure and heart rate in one patient (example), which was recorded preoperatively; *first panel*: systolic and diastolic blood pressure; *second panel*: heart rate

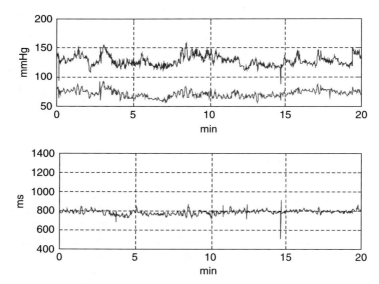

Fig. 3 Tachograms of blood pressure and heart rate in the same patient (Fig. 2), but recorded post-operatively; *first panel*: systolic and diastolic blood pressure; *second panel*: heart rate

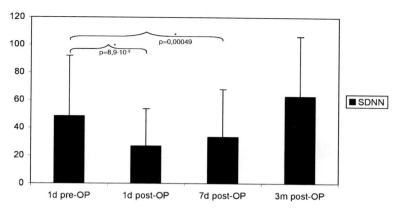

Fig. 4 Heart rate variability: Standard deviation of normal-to-normal beats in ms (SDNN) throughout the postoperative period

A comparably significant decline of heart rate variability was showed for frequency domain parameters 24 h after surgery. Partial recovery could be observed 7 days after surgery for both high frequency (HF) and low frequency components (LF) of beat-to-beat – intervals (Fig. 5).

A similar behaviour was present regarding the baroreflex sensitivity of the parasympathetically mediated portion throughout the postoperative course. 24 h after surgery the bradycardic regulations were depressed slightly more than the strength (Fig. 6). Recovery established earlier in the number than the strength of regulation. Both regulation properties were significantly increased three months after the operation as compared to the analysis after one week and had returned to normal preoperative levels by that time.

Number of regulations and the strength of control revealed an almost parallel behaviour throughout the postoperative course.

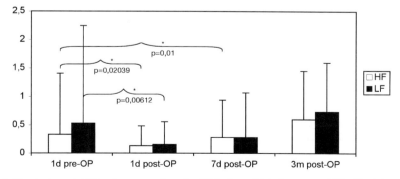

Fig. 5 Heart rate variability: frequency domain. *White bars*: high frequency (HF), *black bars*: low frequency (LF)

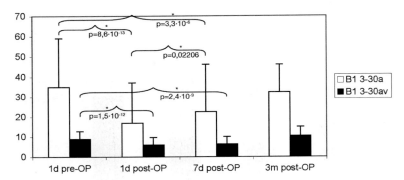

Fig. 6 Baroreflex: bradycardic control. *White bars*: total number of regulations, *black bars*: strength of regulation in ms/mmHg. There is a highly significant reduction of vagally mediated baroreflex control until one week after surgery, which is completely reversible after three months

In the analysis of the tachycardic response of the baroreflex, the delayed mode (shift_3, for explanation see methods section) was chosen to respect the slower regulation properties of the sympathetic system. The strength of the tachycardic regulation did not show any tendency towards recovery after one week, which, however, was present after three months. The number of tachycardic regulations showed completely different kinetics throughout the postoperative course. While they appeared unchanged one day after surgery, they showed a significant decrease after one week. Recovery after three months was comparable to the remaining baroreflex parameters (Fig. 7).

As representative parameters of the nonlinear dynamics analyses, WPSUM13 was chosen as a parameter describing increased variability (higher values represent higher variability) and, correspondingly, WPSUM02 representing decreased variability (Fig. 8). Regarding both parameters, there is a decrease of variability caused

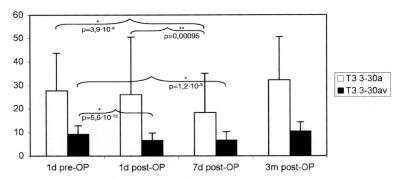

Fig. 7 Baroreflex: tachycardic regulation. *White bars*: total number of regulations; *black bars*: strength of regulation in ms/mmHg. Note the different kinetics of number and strength of regulation

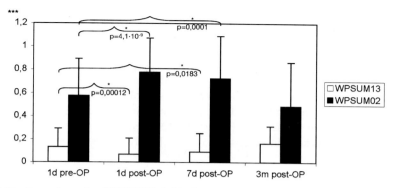

Fig. 8 Nonlinear dynamics: WPSUM13 (*white bars*) and WPSUM02 (*black bars*). Data given as percentage of all word types

by the operation similar to the behaviour of the time domain parameter SDNN. A minimum value can be found at the first postoperative day followed by the tendency towards recovery.

6 Discussion

This prospective study examined the influence of cardiac surgery with heart-lung machine on cardiovascular autonomic control in a consecutive, non-selected cohort of patients undergoing various types of operations, especially emphasizing the changes during the first postoperative month. While the parameters of HRV, nonlinear dynamics and parasympathetically mediated baroreflex response showed a clear tendency to recover after an initial steep decline, the sympathetically mediated baroreflex surprisingly yielded a different behaviour: being more or less unaltered one day after surgery, the number of regulations declined until one week, and recovered again after one month.

The analysis of BRS and HRV provides information about the individual risk in cardiac patients and is significantly altered in these patients as compared to healthy volunteers [1,2]. It was shown in a previous study that heart surgery with ECC leads to a marked alteration of the BRS and HRV, as expressed by time and frequency domain parameters and nonlinear dynamics, respectively [5].

There are several studies investigating the effects of heart surgery on autonomic control, all of them demonstrating impairment immediately after surgery. Two major hypotheses try to explain the early reduction of autonomic function:

1) the effects of anaesthesia and
2) direct surgical damage of efferent cardiac autonomic nerves.

While Niemela and co-workers [10] hardly could observe a tendency to recovery in HRV-parameters after 6 weeks, Brown et al. [4] described restoration after 12 weeks. The authors of the latter paper concluded, that time needed for recovery has to be markedly longer than 6 weeks.

Our data confirm the early loss of autonomic function described by others and the restoration taking place after three months. There may be a role of anesthesia for these alterations, however, the observed persistence until one week after surgery suggests additional influences like trauma, cardioplegic arrest or scar formation [11]. This opinion is in concordance with the findings of earlier trials, desribing absent or only minor influences of anesthesia or medications such as ß-blockers, amiodarone, Ca-blockers, inotropes and nitrates [10, 12–14].

As compared to earlier works, we found one marked difference concerning the course of recovery: there was a clear discrepancy between vagal and sympathetic components of the baroreflex, indicating different modes of influence on the two portions of the autonomic system and thus emphasizing the necessity to include parameters describing regulation mechanisms from different points of view.

Recovery of autonomic function probably is strongly related to the way operations and operative management are performed, therefore discrepancies in studies done in different institutions can hardly be analysed. Direct injury to the surface of the heart and the great vessels is more pronounced in patients undergoing valve surgery as compared to CABG-surgery, therefore faster recovery may be expected in patients with isolated coronary surgery. In contrast to other studies, a significant proportion of the patients in our trial had valve surgery, but still recovery seemed to take place faster. Therefore complete transsection of nerve fibres is unlikely, because, as demonstrated in heart transplant studies [15], complete reinnervation is rare, and, if present, needs more than three months. If a major direct trauma shall be discussed, a temporal effect of crushing, piercing or stretching injuries, which is reversible in a relatively short period of time seems most likely in our patients.

Analysing baroreflex response, there was a surprising difference between bradycardic and tachycardic regulation. The parasympathetically mediated bradycardic response showed a steep decline 24 h after surgery with a subsequent tendency to recover after one week and complete restoration after three months. The sympathetically mediated tachycardic regulation was not altered or only slightly reduced one day after the operation, but had no tendency to recover or even revealed a further significant decrease one week after the operation. For all parameters of the tachycardic regulation, full restoration was present after three months. So in contrast to the vagal portion, the sympathetically mediated part of the autonomic system seems to be more severely altered by changes taking place after the operation as compared to intraoperative adverse effects. A major influence of medication given in the postoperative period again can most probably be ruled out, because there was no striking difference in drugs applied at day 7 or the third month.

So the hypothesis remaining for explanation from a pathophysiologic point of view is the following: scar formation accompanied by an inflammatory reaction of the surface of the heart and the pericardium seems to predominantly affect the sympathetic fibres and subsequently impair tachycardic regulation.

Our studies show, that the postoperative course of patients undergoing heart surgery is characterised by an imbalance of the autonomic system persisting for at least one week. From an abundance of studies it is known, that within the first postoperative week patients carry the highest risk of cardiac related adverse events like sudden cardiac death, low cardiac output or malignant rhythm disorders [16] and that any imbalance of autonomic regulation predisposes to complications [17]. Although a causative relationship cannot be proved at the present state of knowledge, the effects of deranged autonomic control on the clinical course of patients with cardiovascular diseases has been demonstrated in earlier experimental and clinical studies [2, 18].

In that light, careful description of the changes taking place over time can be considered the first step to gain deeper insight into the underlying pathology in order to pave the way to establish models of risk stratification and knowledge-based therapy able to predict the risk of an individual patient, thus complementing the information obtained by established scoring systems describing cohorts of patients [19].

It can be summarized, that autonomic control is altered by cardiac surgery and has the potential to recover within a short period of time. Further studies now have to focus on the discrimination of patients at high operative or postoperative risk according to their specific patterns of the dynamics of autonomic control for preoperative risk stratification and postoperative monitoring.

References

1. D. Wichterle, J. Simek, M.T. La Rovere, P.J. Schwartz, A.J. Camm, M. Malik, *Prevalent low-frequency oscillation of heart rate: novel predictor of mortality after myocardial infarction*, Circulation **110**, 1183–1190 (2004)
2. M. T. La Rovere, G. D. Pinna, S. H. Hohnloser, F. I. Marcus, A. Mortara, R. Nohara, J. T. Bigger Jr, A. J. Camm, P. J. Schwartz, *ATRAMI investigators. Autonomic tone and reflexes after myocardial infarction. Baroreflex sensitivity and heart rate variability in the identification of patients at risk for life-threatening arrhythmias: implications for clinical trials*. Circulation **103**, 2072–2077 (2001)
3. H. Malberg, R. Bauernschmitt, U. Meyerfeldt, A. Schirdewan, N. Wessel, *Short-term heart rate turbulence analysis versus variability and baroreceptor sensitivity in patients with dilated cardiomyopathy*. Z. Kardiol. **92**, 547–557 (2003) translation in: Ind. Pacing Electrophysiol. J. **4**, 162–175 (2004)
4. C. A. Brown, L. A. Wolfe, G. Hains, G. Ropchan, J. Parlow, *Spontaneous baroreflex sensitivity after coronary artery bypass graft surgery as a function of gender and age*. Can. J. Physiol. Pharmacol. **9**, 894–902 (2003)
5. R. Bauernschmitt, H. Malberg, N. Wessel, B. Kopp, E. U. Schirmbeck, R. Lange, *Impairment of cardiovascular autonomic control in patients early after cardiac surgery*. Eur. J. Cardiothorac. Surg. **25**(3), 320–326 (2004)
6. H. Malberg, N. Wessel, A. Hasart, K. J. Osterziel, A. Voss, *Advanced analysis of the spontaneous baroreflex sensitivity, blood pressure and heart rate variability in patients with dilated cardiomyopathy*. Clin. Sci. **102**, 465–473 (2002)
7. *Heart rate variability: Standards of measurement, physiological interpretation, and clinical use. Task Force of The European Society of Cardiology and The North American Society of Pacing and Electrophysiology*. European Heart Journal **17**, 354–381(1996)
8. J. Kurths, A. Voss, A. Witt, P. Saparin, H. J. Kleiner, N. Wessel, *Quantitative analysis of heart rate variability*. Chaos **5**, 88–94 (1995)

9. A. Voss, J. Kurths, H. J. Kleiner, A. Witt, N. Wessel, P. Saparin, K. K. Osterziel, R. Schurath, R. Dietz, *The application of methods of non-linear dynamics for the improved and predictive recognition of patients threatened by sudden cardiac death.* Cardiovasc. Res. **31**, 419–433 (1996)
10. M. J. Niemela, K. E .J. Airaksinen, K. U. O. Tahvanainen, M. K. Linnaluoto, J. T. Takkunen, *Effect of coronary artery bypass grafting on cardiac parasympathetic nervous function.* Eur. Heart J. **13**, 932–935 (1992)
11. Z. K. Wu, S. Vikman, J. Laurikka, E. Pehkonen, T. Iivainen, H. V. Huikuri, M. R. Tarkka, *Nonlinear heart rate variability in CABG patients and the preconditioning effect.* Eur. J. Cardiothorac. Surg. **28**(1), 109–113 (2005)
12. T. T. Laitio, H. V. Huikuri, E. S. Kentala, T. H. Makikallio, J. R. Jalonen, H. Helenius, K. Sariola-Heinonen, S. Yli-Mayry, H. Scheinin, *Correlation properties and complexity of perioperative RR-interval dynamics in coronary artery bypass surgery patients.* Anesthesiology **93**, 69–80 (2000)
13. C. W. jr Hogue, P. K. Stein, I. Apostolidou, D. G. Lappas, R. E. Kleiger, *Alterations in temporal patterns of heart rate variability after coronary artery bypass graft surgery.* Anesthesiology **81**, 1356–1364 (1994)
14. C. D. Kuo, G. Y. Chen, S. T. Lai, Y. Y. Wang, C. C. Shih, J. H. Wang, *Sequential changes in heart rate variability after coronary artery bypass grafting.* Am. J. Cardiol. **83**, 776–779 (1999)
15. F. Beckers, D. Ramaekers, G. Speijer, H. Ector, J. Vanhaecke, J. B. Verheyden, J. Van Cleemput, W. F. Droogne, F. Van de Werf, A. E. Aubert, *Different evolutions in heart rate variability after heart transplantation: 10-year follow-up.* Transplantation **10**, 1523–1531 (2004)
16. P. K. Smith, M. Carrier, J. C. Chen, A. Haverich, J. H. Levy, P. Menasche, S. K. Shernan, F. Van de Werf, P. X. Adams, T. G. Todaro, E. Verrier, *Effect of pexelizumab in coronary artery bypass graft surgery with extended aortic cross-clamp time.* Ann. Thorac. Surg. **82**(3), 781–788 (2006)
17. N. Wessel, Ch. Ziehmann, J. Kurths, U. Meyerfeldt, A. Schirdewan, A. Voss, *Short-term forecasting of life-threatening cardiac arrhythmias based on symbolic dynamics and finite-time growth rate.* Phys. Rev. E **61**, 733–739 (2000)
18. A. Elan, D. P. Zipes, *Right ventricular infarction causes heterogeneous autonomic denervation of the viable peri-infarct area.* Circulation **97**(5), 484–492 (1998)
19. F. Biancari, O. P. Kangasniemi, J. Luukkonen, S. Vuorisalo, J. Satta, R. Pokela, T. Juvonen, *EuroSCORE predicts immediate and late outcome after coronary artery bypass surgery.* Ann. Thorac. Surg. **82**(1), 57–61 (2006)

Application of Empirical Mode Decomposition to Cardiorespiratory Synchronization

Ming-Chya Wu and Chin-Kun Hu

Abstract A scheme based on the empirical mode decomposition (EMD) and synchrogram introduced by Wu and Hu [Phys. Rev. E **74**, 051917 (2006)] to study cardiorespiratory synchronization is reviewed. In the scheme, an experimental respiratory signal is decomposed into a set of intrinsic mode functions (IMFs), and one of these IMFs is selected as a respiratory rhythm to construct the cardiorespiratory synchrogram incorporating with heartbeat data. The analysis of 20 data sets from ten young (21–34 years old) and ten elderly (68–81 years old) rigorously screened healthy subjects shows that regularity of respiratory signals plays a dominant role in cardiorespiratory synchronization.

Keywords Empirical mode decomposition · Intrinsic mode functions · Cardiorespiratory synchrogram

1 Introduction

Physiological systems are nonlinear, and biomedical signals are apparently random or aperiodic in time. These systems can serve as a playground for the study of analysis techniques of nonlinear dynamics. Recently, the study of oscillations and couplings in these systems has gained increasing attention [1–19]. Among these, the nature of the couplings between human cardiovascular and respiratory systems has been widely studied [18–26], and is known to be both neurological [1] and mechanical [2]. The interactions between the two systems result in the well-known modulation of heart rates, known as respiratory sinus arrhythmia (RSA). Recent studies suggest that beside modulations, there is also synchronization between them.

Almasi and Schmitt reported that there are voluntary synchronization between subjects' breathing and cardiac cycle [3], in which subjects, signaled by a tone derived from the electrocardiograms (ECGs), inspired for a fixed number of heart

---

M.-C. Wu (✉)
Research Center for Adaptive Data Analysis, National Central University, Chungli 32001, Taiwan
e-mail: mcwu@ncu.edu.tw

beats followed by expiration for a fixed number of heart beats [3]. Recently, Schäfer et al. [5, 6] and Rosenblum et al. [7] applied the concept of phase synchronization of chaotic oscillators [15] to analyze irregular non-stationary bivariate data from cardiovascular and respiratory systems, and introduced the cardiorespiratory synchrogram (CRS) to detect different synchronous states and transitions between them. They found sufficient long period of synchronization and concluded that the cardiorespiratory synchronization and RSA are two competing factors in cardiorespiratory interactions. Latter, Tolddo et al. [8] found that synchronization was less abundant in normal subjects than in the transplant patients, which indicated that the physiological condition of the latter promotes cardiorespiratory synchronization. More recently, Kotani et al. [14] developed a physiologically model to study the phenomena, and showed that both the influence of respiration on heartbeat and the influence of heartbeat on respiration are important for cardiorespiratory synchronization.

Up to now, cardiorespiratory synchronization has been reported in young health athletes [5, 6], healthy adults [9–11], heart transplant patients [9], infants [12], and anesthetized rats [13]. Since the studies are based on measured data, the data processing method plays a crucial role in the outcome. An essential task for the studies is to process such signals and pickup essential component(s) from experimental respiratory signals dressing with noise. Except for the Fourier spectral analysis which has been widely used, to date there have been several approaches to preprocess real data for this purpose [27–33]. Most of these approaches require that the original time series should be stationary and/or linear, while respiratory signals are noisy, nonlinear, and non-stationary. As a result, a number of filters may be used to filter out noises from real data, while the capabilities and effectiveness of the filtration are usually questionable. There is also no strict criterion to judge what is the inherent dynamics and what is contribution of the external factors and noise in measured data. Improper approaches might lead to misleading results.

To overcome above difficulties, Wu and Hu [26] suggest using the empirical mode decomposition (EMD) method proposed by Huang et al. [34] and the Hilbert spectral analysis,[1] as a candidate for such studies. Unlike conventional filters, the EMD provides an effective way to extract respiratory rhythms from experimental respiratory signals. The EMD uses the sifting process to eliminate riding waves and make the wave-profiles more symmetric. The expansion of the turbulence data set in EMD has only a finite-number of locally non-overlapping time scale components, known as intrinsic mode functions (IMFs). These IMFs form a complete set and are orthogonal to each other. The adaptive properties of EMD to empirical data make it easy to give physical significations to IMFs, and allow us to choose a certain IMF as a respiratory rhythm [26]. As an IMF is selected for the respiratory rhythm, one can further use CRS to detect synchronization. In this article, we will review the scheme proposed by Wu and Hu [26], and focus on the application of EMD

---

[1] Besides the Hilbert spectral analysis, one can also use other methods to process the data obtained from EMD, see e.g. Refs. [17, 18].

to cardiorespiratory synchronization. Details of the study will be referred to their original paper [26].

This article is organized as follows. In Section 2, we introduce the EMD method. In Section 3, the EMD is used to extract the respiratory rhythm from experimental data and the Hilbert transform is used to calculate the instantaneous phase of the respiratory time series. The CRS is then constructed by assessing heartbeat data on the phase of the respiratory signal, and is used to visually detect the epochs of synchronization in Section 4. In Section 5, we investigate the correlation between regularity of respiratory signals and cardiorespiratory synchronization. Finally, we discuss our results in Section 6.

2 Empirical Mode Decomposition

The EMD is an empirically based data-analysis method. It was developed from the assumption that any data consists of different simple intrinsic modes of oscillations. The essence of the EMD is to identify the intrinsic oscillatory modes by characteristic time scales in the data empirically, and then decompose the data accordingly [34]. This is achieved by "sifting" data to generate IMFs. The IMFs obtained by the EMD are a set of well-behaved intrinsic modes and are symmetric with respect to the local mean and have the same numbers of zero crossings and extrema. The algorithm to create IMFs in the EMD has two main steps [26, 34]:

Step-1: Identify local extrema in the experimental data $\{x(t)\}$. All the local maxima are connected by a cubic spline line $U(t)$, which forms the upper envelope of the data. Repeat the same procedure for the local minima to produce the lower envelope $L(t)$. Both envelopes will cover all the data between them. The mean of upper envelope and lower envelope $m_1(t)$ is given by:

$$m_1(t) = \frac{U(t) + L(t)}{2}. \qquad (1)$$

Subtracting the running mean $m_1(t)$ from the original time series $x(t)$, we get the first component $h_1(t)$,

$$h_1(t) = x(t) - m_1(t). \qquad (2)$$

The resulting component $h_1(t)$ is an IMF if it is symmetric and have all maxima positive and all minima negative. An additional condition of intermittence can be imposed here to sift out waveforms with certain range of intermittence for physical consideration. If $h_1(t)$ is not an IMF, the sifting process has to be repeated as many times as it is required to reduce the extracted signal to an IMF. In the subsequent sifting process steps, $h_1(t)$ is treated as the data to repeat steps mentioned above,

$$h_{11}(t) = h_1(t) - m_{11}(t). \qquad (3)$$

Again, if the function $h_{11}(t)$ does not yet satisfy criteria for IMF, the sifting process continues up to k times until some acceptable tolerance is reached:

$$h_{1k}(t) = h_{1(k-1)}(t) - m_{1k}(t). \tag{4}$$

Step-2: If the resulting time series is an IMF, it is designated as $c_1 = h_{1k}(t)$. The first IMF is then subtracted from the original data, and the difference r_1 given by

$$r_1(t) = x(t) - c_1(t). \tag{5}$$

is the residue. The residue $r_1(t)$ is taken as if it were the original data, and we apply to it again the sifting process of *Step-1*.

Following the procedures of *Step-1* and *Step-2*, we continue the process to find more intrinsic modes c_i until the last one. The final residue will be a constant or a monotonic function which represents the general trend of the time series. Finally, we obtain

$$x(t) = \sum_{i=1}^{n} c_i(t) + r_n, \tag{6}$$

$$r_{i-1}(t) - c_i(t) = r_i(t). \tag{7}$$

The instantaneous phase of IMF can be calculated by applying the Hilbert transform to each IMF, say the rth component $c_r(t)$. The procedures of the Hilbert transform consist of calculation of the conjugate pair of $c_r(t)$, i.e.,

$$y_r(t) = \frac{1}{\pi} P \int_{-\infty}^{\infty} \frac{c_r(t')}{t - t'} dt', \tag{8}$$

where P indicates the Cauchy principal value. With this definition, two functions $c_r(t)$ and $y_r(t)$ forming a complex conjugate pair, define an analytic signal $z_r(t)$:

$$z_r(t) = c_r(t) + i y_r(t) \equiv A_r(t) e^{i\phi_r(t)}, \tag{9}$$

with amplitude $A_r(t)$ and the instantaneous phase $\phi_r(t)$ defined by

$$A_r(t) = \left[c_r^2(t) + y_r^2(t)\right]^{1/2}, \tag{10}$$

$$\phi_r(t) = \tan^{-1}\left(\frac{y_r(t)}{c_r(t)}\right). \tag{11}$$

3 Data Acquisition and Processing

The empirical data consisting of 20 data sets were collected by the Harvard medical school in 1994 [35]. Ten young (21–34 years old) and ten elderly (68–81 years old) rigorously-screened healthy subjects underwent 120 minutes of continuous supine resting while continuous ECG and respiration signals were collected. The continuous ECG and respiration data were digitized at 250 Hz (respiratory signals were latter preprocessed to be at 5 Hz). Each heartbeat was annotated using an automated arrhythmia detection algorithm, and each beat annotation was verified by visual inspection. Among these, records f1y01, f1y02, ..., f1y10 were obtained from the young cohort, and records f1o01, f1o02, ..., f1o10 were obtained from the elder cohort. Each group of subjects includes equal numbers of men and women.

The respiratory signals represent measures of the volume of expansion of ribcage, so the corresponding data are all positive numbers and there are no zero crossings. In addition to respiratory rhythms, the data also contain noises originating from measurements, external disturbances and other factors. In this work we apply the EMD [34] to preprocess the data. From to the decomposition of EMD, one can select one component as the respiratory rhythm according to the criteria of intermittencies of IMFs imposed in *Step-1* as an additional sifting condition [26]. Note that among IMFs, the first IMF has the highest oscillatory frequency, and the relation of intermittence between different modes is $\tau_n = 2^{n-1}\tau_1$ with τ_n the intermittence of the nth mode. More explicitly, the procedures of data processing are as follows. (i) Apply EMD to decompose the data into several IMFs. The decomposition acquires input of the criterion of intermittence as the parameters in the sifting process, and we use the time scale of a respiratory cycle as the criteria. Since the respiratory signal was preprocessed to a sampling rate of 5 Hz, there are (10–30) data points in one cycle.[2] Then, for example, we can use c_1: (3–6), c_2: (6–12), c_3: (12–24), etc. After the sifting processes of EMD, the original respiratory data is decomposed into n empirical modes $c_1, c_2, \ldots, c_n$, and a residue r_n. (ii) Visually inspect the resulting IMFs. If the amplitude of certain mode is dominant and the wave-form is well distributed, the data is said to be well decomposed and the decomposition is successfully completed. Otherwise, the decomposition may be inappropriate, and we have to repeat step (i) with different parameters.

Figure 1 shows the decomposition of an empirical signal with a criterion of the intermittence being (3–6) data points for c_1, and ($3 \times 2^{n-1}$–3×2^n) data points for c_n's with $n > 1$. Comparing $x(t)$ with c_i's, it is obvious that c_3 preserves the main structure of the signal and is dominant in the decomposition. We thus pickup the third component c_3, corresponding to (12–24) data points per respiratory cycle, as

---

[2] The number of breathing per minute is about 18 for adults, and about 26 for children. For different healthy states, the number of respiratory cycles may vary case by case. To include most of these possibilities, we take respiratory cycles ranging from 10 to 30 times per minute, and each respiratory cycle then roughly takes 2–6 seconds, i.e., (10–30) data points.

the respiratory rhythm. After one of IMFs is selected as the respiratory rhythm, we can proceed in the next step to construct CRS.

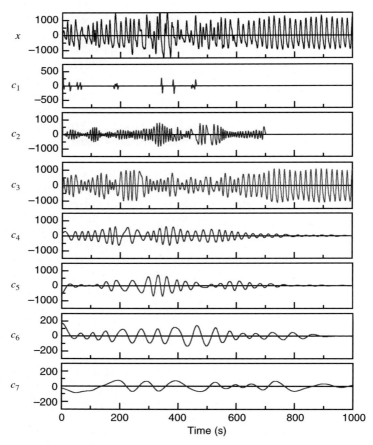

Fig. 1 Example of EMD for a typical respiratory time series data (flo01). The criteria for intermittence in the sifting process is (3–6) data points per cycle for c_1. Signal $x(t)$ is decomposed into 14 components including 13 IMFs and 1 residue. Here only the first 7 components are shown. After Ref. [26]

4 Cardiorespiratory Synchrogram

Cardiorespiratory synchronization is a process of adjustment of rhythms due to interactions between cardiovascular and respiratory systems. These interactions can lead to a perfect locking of their phases, whereas their amplitudes remain chaotic and non-correlated [4]. If the phases of respiratory signal ϕ_r and heartbeat ϕ_c are coupled in a fashion that a cardiovascular system completes n heartbeats in m

respiratory cycles, then a roughly fixed relation can be proposed. In general, there is a phase and frequency locking condition [4–6]

$$|m\phi_r - n\phi_c| \leq \text{const.},\tag{12}$$

with m, n integer. According to Eq. (12), for the case that ECG completes n cycles while the respiration completes m cycles, it is said to be synchronization of n cardiac cycles with m respiratory cycles. Using the heartbeat event time t_k as the time frame, Eq. (12) implies the relation

$$\phi_r(t_{k+m}) - \phi_r(t_k) = 2\pi m.\tag{13}$$

Furthermore, by defining

$$\Psi_m(t_k) = \frac{1}{2\pi}[\phi_r(t_k) \bmod 2\pi m]\tag{14}$$

and plotting $\psi_m(t_k)$ versus t_k, synchronization will result in n horizontal lines in case of n:m synchronization. By choosing n adequately, a CRS can be developed for detecting the synchronization between heartbeat and respiration [5,6].

Example of 3:1 synchronization with $n = 6$ and $m = 2$ is shown in Fig. 2a, where phase locking appear in several epochs, e.g. at 2800–3600s, and there are also frequency locking, e.g. at 400s, near which there are n parallel lines with the same positive slope. For comparison, we also show the results of the same subject in 1800–3600s, but with respiratory signals without filtering, preprocessed by the standard filters and the EMD in Fig. 2b. The windows of the standard filters are

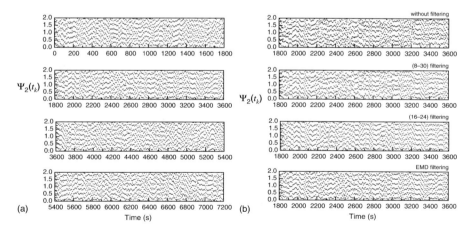

Fig. 2 Cardiorespiratory synchrogram for a typical subject (f1o06). (**a**) Empirical data are preprocessed by the EMD. There are about 800s synchronization at 2800–3600s, and several spells of 50–300s at other time intervals. (**b**) Comparison of the results without filtering (*top one*), preprocessed by the standard filters with windows of (8–30) and (16–24) cycles per minute (the second and the third ones), and the EMD method (*bottom one*). After Ref. [26]

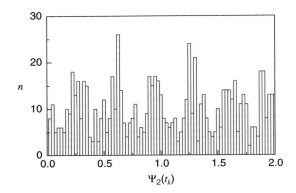

Fig. 3 Histogram of phase for the phase locking period from 2800s to 3600s for a typical subject (flo06) shown in Fig. 2a. After Ref. [26]

(8–30) and (16–24) cycles per min. In general, some noise dressed signals can still show synchronization in some epochs but the Hilbert spectral analysis failed at some time intervals (e.g., around 3400–3600s of the case without filtering), and over-filtered signals reveal too strong synchronization (filter with window of 16–24). In other words, global frequency used in standard filters may dissolve local structures of the empirical data. This does not happen in the EMD filtering.

Figure 3 shows the histogram of phases for the phase locking period from 2800 to 3600s in Fig. 2a. Significant higher distribution can be found at $\psi_2 \approx 0.25, 0.6, 0.9, 1.25, 1.6, 1.9$ in the unit of 2π, indicating heartbeat events occur roughly at these respiratory phase during this period. Following above procedures, we analyze data of 20 subjects, and the results are summarized in Table 1. The results are ordered by

Table 1 Summary of our results. 20 subjects are ordered by the strength (total time length) of the cardiorespiratory synchronization. After Ref. [26]

Code	Sex	Age	Synchronization
flo06	F	74	3:1(800s, 300s, 250s, 150s, 100s, 50s)
fly05	M	23	3:1(350s, 300s, 200s, 100s)
flo03	M	73	3:1(200s, 50s, 30s)
fly10	F	21	7:2(200s, 50s), 4:1(50s)
flo07	M	68	7:2(120s, 100s, 80s)
flo02	F	73	3:1(100s, several spells of 50s)
fly01	F	23	7:2(several spells of 30s)
fly04	M	31	5:2(80s, 50s, 30s)
flo08	F	73	3:1(50s, 30s)
fly06	M	30	4:1(50s, 30s)
flo01	F	77	7:2(several spells of 50s)
fly02	F	28	3:1(50s)
fly08	F	30	3:1(50s)
flo10	F	71	3:1(30s)
flo05	M	76	No synchronization detectable
fly07	M	21	No synchronization detectable
fly09	F	32	No synchronization detectable
fly03	M	34	No synchronization detectable
flo09	M	71	No synchronization detectable
flo04	M	81	No synchronization detectable

the strength of the cardiorespiratory synchronization. From our results, we do not find specific relations between the occurrence of synchronization and sex of the subjects as in Refs. [5,6]. Here we note that if we use other filters to the same empirical data, we will have different results depending on the strength of synchronization. Wu and Hu [26] found that from the aspect of data processing that could preserve the essential features of original empirical data, the EMD approach is better than Fourier-based filtering.

5 Correlation and Regularity

As noted above, data processing method plays a crucial role in the analysis of real data. Over-filtered respiratory signals may lose detailed structures and become too regular. It follows that final conclusions are methodological dependent. One might then ask how the results depend on the data processing methods. This problem arises when one addresses the issue of existence or strength of the cardiorespiratory synchronization, and the answers may be helpful for understanding the mechanisms of the synchronization.

In general, the existence of cardiorespiratory synchronization is confirmed simply when it is observed in enough subjects analyzed by various approaches. The existing studies have positive answers on its existence [5, 6, 9–13]. Nevertheless, the strength of synchronization for these subjects may depend on the methods used, and need further investigations. For this purpose, we test the correlations between cardiac and respiratory signals as well as their regularities. We first consider two data sets: (f1o06.res, f1o06.hrt) and (f1y05.res, f1y05.hrt). Here notation "code.signal" indicates one code and its corresponding signal. Both of these two data sets, f1o06 and f1y05, have been analyzed to show 3:1 synchronization in some periods. The synchronization exhibited by these two data sets in an interval from 2000s to 3600s is shown respectively in Fig. 4a and b. We interchange the respiratory and cardiac time series of them to be (f1o06.res, f1y05.hrt) and (f1y05.res, f1o06.hrt), and then construct their synchrograms. The results are shown in Fig. 4c and d, respectively. There are still phase locking appearing in shorter spells for the "mixed" data, such as at 3000s of Fig. 4c and at 2000s of Fig. 4d. This implies the synchronization should be detectable provided that there are characteristic features coupled between respiratory and cardiac signals. Therefore, emergence of short shells of synchronization does not necessarily imply true coupling between cardiovascular and respiratory systems. If cardiorespiratory synchronization exists in a subject, the cardiovascular and respiratory systems must correlate with the same variation scheme of intermittence such that synchronization can appear again and again in some time intervals. Hence, the phase locking in the synchrogram of Fig. 2a, where synchronization disappears and recovers repeatedly at 1800–3600s due to the variation of intermittence of respiratory time series indicates true cardiorespiratory synchronization.

Next, we test the dependence of the results on the regularity of signals. In our study, cardiac signals are regular enough [26], which implies the regularity of

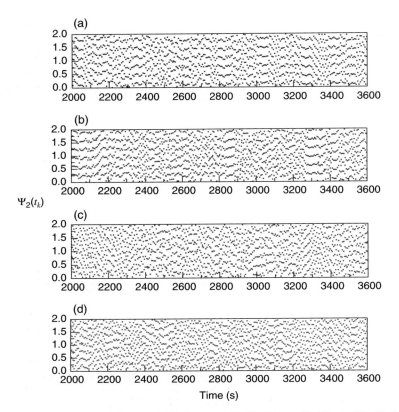

Fig. 4 Cardiorespiratory synchrogram for data sets (**a**) (f1o06.res, f1o06.hrt), (**b**) (f1y05.res, f1y05.hrt), (**c**) (f1o06.res, f1y05.hrt), (**d**) (f1y05.res, f1o06.hrt), at period 2000–3600s. After Ref. [26]

cardiac signals is not necessarily related to the strength of synchronization. In contrast to cardiac signals, real respiratory signals are essentially irregular. Here we will not measure regularity of respiratory cycles directly, but compare synchronization in CRS for various sets of cardiac and respiratory time series. We introduce an artificial respiratory signal generated by a generic cosine wave $s(S_0, T, t)$,

$$s(S_0, T, t) = S_0 \cos\left(\frac{2\pi t}{T}\right), \tag{15}$$

where S_0 is the amplitude and T is the period. The frequency of this wave is fixed and the phase varies regularly. We first construct the synchrogram for $[s(S_0, T, t),$ f1o06.hrt]. The results are shown in Fig. 5, in which different periods $T = 15$, 16, 17, 17.6, 18 have been used. According to Fig. 5, the cardiac signals for this subject are rather regular, and a fixed heartbeat frequency can last relatively long time, even if it changes finally. When T is a multiple of 3, such as $T = 15$ and

Application of Empirical Mode Decomposition

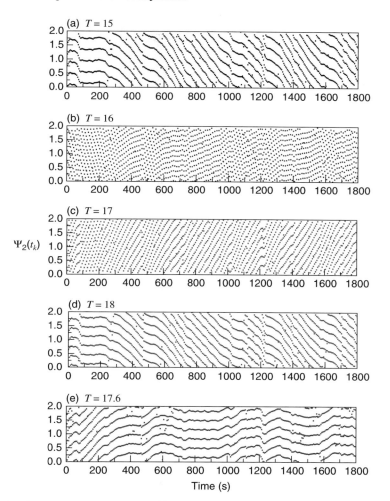

Fig. 5 Cardiorespiratory synchrogram for data sets $[s(S_0 = 1000, T, t), \text{f1o06.hrt}]$ with (**a**) $T = 15$, (**b**) $T = 16$, (**c**) $T = 17$, (**d**) $T = 18$, and (**e**) $T = 17.6$. After Ref. [26]

$T = 18$, there are synchronization spells observed at the period from 100s to 220s, and phase locking at the other epochs. For $T = 17.6$, phase locking can be observed at most epochs of the period. Here we should note that, comparing Figs. 2(a) and 5, a short spell from 100 s to 220s appears as phase locking corresponds to respiratory intermittence $T = 18$. However, a short spell from 1220s to 1350s corresponds to respiratory intermittence roughly about $T = 17.6$. Even the intermittence varies, the synchronization persists. This indicates the existence of correlations.

Comparing the patterns of the periods where synchronization occurs in Fig. 5 and the corresponding periods in Fig. 4, we find that cardiac signals are regular enough such that synchronization occurs at the framework of regular time series,

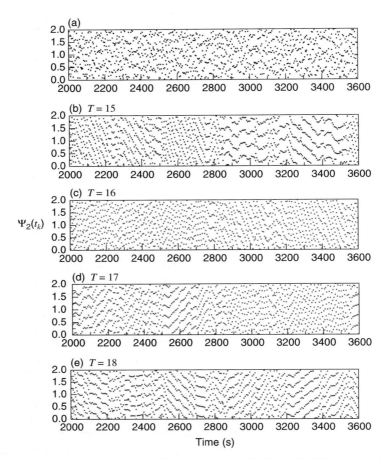

Fig. 6 Cardiorespiratory synchrogram for data set (**a**) (f1o09.res, f1o09.hrt), and data sets $[s(S_0 = 1000, T, t), \text{f1o09.hrt}]$, with (**b**) $T = 15$, (**c**) $T = 16$, (**d**) $T = 17$, and (**e**) $T = 18$. After Ref. [26]

and respiratory cycles are not regular enough such that there is weaker or even no synchronization in the corresponding periods. To have more results for comparison, we examine another subject having data set (f1o09.res, f1o09.hrt), which has no synchronization at all in the preceding analysis. The synchrogram for data set (f1o09.res, f1o09.hrt) is shown in Fig. 6a, and data sets $[s(S_0, T, t), \text{f1o09.hrt}]$ with $T = 15, 16, 17, 18$ are respectively shown in Fig. 6b–e. We find that there are short spells phase locking or frequency locking appear, and the length of the spells depend on the period T. Therefore, more regular respiratory signals have better manifestation of synchronization.

From above investigation, we conclude that: (i) In most cases, cardiac oscillations are more regular than respiratory oscillations and the respiratory signal is the key factor for the strength of the cardiorespiratory synchronization. (ii) Cardiorespiratory phase locking and frequency locking take place when respiratory oscillations become regular enough and have a particular frequency relation coupling with cardiac oscillations. We observed the intermittence of respiratory oscillation varies with time but synchronization persists in some subjects, such as codes f1o06 and f1y05. This confirms correlations in the cardiorespiratory synchronization. (iii) Over-filtered respiratory signals may be too regular, and in turn, appear to have stronger synchronization than they shall have. Therefore, if the Fourier based approach with narrow band filtration is used, some epochs of phase locking or frequency locking should be considered as being originated from these effects.

6 Discussion

We have reviewed the scheme introduced by Wu and Hu [26] to study cardiorespiratory synchronization. The scheme is based on the EMD and CRS. The advantage of using the EMD is that it catches primary structures of respiratory rhythms based on its adaptive feature [34, 36]. By imposing the intermittency criteria based on physiological condition revealing from empirical time series, this feature allows us to effectively keep the signal structures and avoid the introduction of artificial signals easily appear in the Fourier-based filters with priori bases [26]. Furthermore, the introduction of IMFs in EMD provides a reasonable definition of instantaneous phase. This advantage is helpful for drawing reliable conclusions on the studies of empirical data. The study supports the existence of the cardiorespiratory synchronization. However, no difference in synchronization between two different age groups and two different sex groups were found. At the current stage, even cardiorespiratory synchronization has been observed in a number of studies, there is still no confident conclusion on its dependence on sex and age due to few subjects were studied and most of them were performed in different physiological stages. Furthermore, the statistics of our results indicates that most synchronization exhibits 3:1 (8 subjects), 4:1 (1 subjects) and 7:2 (4 subjects) synchronization, which is consistent with the report of Ref. [12] that mature physiological subjects (adults) have larger probabilities of 3:1, 4:1, and 7:2 synchronization than 5:2 synchronization.

From a physiological viewpoint, it is difficult to precisely identify the mechanisms responsible for the observed non-linear interactions. From our studies, we found that cardiac oscillations are more regular than respiratory oscillations, and cardiorespiratory synchronization occurs at the period when respiratory signals become regular enough. In other words, the regularity of respiratory signals contributes dominantly to the synchronization. Cardiorespiratory synchronization and RSA are two competing factors in the cardiorespiratory interactions. This observation is consistent with the results reported in Refs. [21, 37].

Finally, it should be remarked that the technique used in this work can also be applied to the analysis of other time series, such as financial time series [38–40]. It is also interesting to extend the technique to analyze signals from many-body systems and study their synchronization.

Acknowledgments This work was supported by the National Science Council of the Republic of China (Taiwan) under Grant No. NSC 95-2112-M-001-008 and National Center for Theoretical Sciences in Taiwan.

References

1. A.C. Guyton, *Textbook of medical physiology*, 8th ed. (Saunders, Philadelphia 1991).
2. L. Bernardi, F. Salvucci, R. Suardi, P.L. Solda, A. Calciati, S. Perlini, C. Falcone, and L. Ricciardi, *Evidence for an intrinsic mechanism regulating heart-rate-variability in the transplanted and the intact heart during submaximal dynamic exercise*, Cardiovasc. Res. **24**, 969–981 (1990).
3. J. Almasi and O.H. Schmitt, *Basic technology of voluntary cardiorespiratory synchronization in electrocardiology*, IEEE Trans. Biomed. Eng. **21**, 264–273 (1974).
4. P. Tass, M.G. Rosenblum, J. Weule, J. Kurths, A. Pikovsky, J. Volkmann, A. Schnitzler, and H.-J. Freund, *Detection of n:m Phase Locking from Noisy Data: Application to Magnetoencephalography*, Phys. Rev. Lett. **31**, 3291 (1998).
5. C. Schäfer, M.G. Rosenblum, J. Kurths, and H.-H. Abel, *Heartbeat synchronized with ventilation*, Nature (London) **392**, 239–240 (1998).
6. C. Schäfer, M.G. Rosenblum, H.-H. Abel, and J. Kurths, *Synchronization in the human cardiorespiratory system*, Phys. Rev. E **60**, 857 (1999).
7. M.G. Rosenblum, J. Kurths, A. Pikovsky, C. Schäfer, P. Tass, H.-H. Abel, *Synchronization in noisy systems and cardiorespiratory interaction*, IEEE Eng. Med. Biol. Mag. **17**, 46–53 (1998).
8. E. Toledo, S. Akselrod, I. Pinhas, and D. Aravot, *Does synchronization reflect a true interaction in the cardiorespiratory system?* Med. Eng. Phys. **24**, 45–52 (2002).
9. E. Toledo, M.G. Roseblum, C. Schäfer, J. Kurhts, and S. Akselrod, *Quantification of cardiorespiratory synchronization in normal and heart transplant subjects*. In: Proc. of Int. Symposium on Nonlinear Theory and its Applications, vol. 1, Lausanne, Presses polytechniques et universitaires romandes, pp. 171–174 (1998).
10. E. Toledo, M.G. Rosenblum, J. Kurths, and S. Akselrod, *Cardiorespiratory synchronization: is it a real phenomenon ?* In: Computers in Cardiology, vol. 26, Los Alamitos (CA), IEEE Computer Society, pp. 237–240 (1999).
11. M.B. Lotric and A. Stefanovska, *Synchronization and modulation in the human cardiorespiratory system*, Physica A **283**, 451–461 (2000).
12. R. Mrowka and A. Patzak, *Quantitative analysis of cardiorespiratory synchronization in infants*, Int. J. Bifurcation and Chaos **10**, 2479–2488 (2000).
13. A. Stefanovska, H. Haken, P.V.E. McClintock, M. Hozic, F. Bajrovic, and S. Ribaric, *Reversible transitions between synchronization states of the cardiorespiratory system*, Phys. Rev. Lett. **85**, 4831 (2000).
14. K. Kotani, K. Takamasu, Y. Ashkenazy, H.E. Stanley, and Y. Yamamoto, *Model for cardiorespiratory synchronization in humans*, Phys. Rev. E **65**, 051923 (2002).
15. M.G. Rosenblum, A.S. Pikovsky, and J. Kurths, *Phase synchronization of chaotic oscillators*, Phys. Rev. Lett. **76**, 1804 (1996).
16. R. Q. Quiroga, J. Arnhold, and P. Grassberger, *Learning driver-response relationships from synchronization patterns*, Phys. Rev. E **61**, 5142 (2000).
17. R.Q. Quiroga, A. Kraskov, T. Kreuz, and P.P. Grassberger, *Performance of different synchronization measures in real data: A case study on electroencephalographic signals*, Phys. Rev. E **65**, 041903 (2002).
18. M.G. Rosenblum, A.S. Pikovsky, and J. Kurths, *Synchronization approach to analysis of biological systems*, Fluctuation and Noise Lett. **4**, L53–L62 (2004).

19. M.G. Rosenblum and A.S. Pikovsky, *Controlling synchronization in an ensemble of globally coupled oscillators*, Phys. Rev. Lett. **92**, 114102 (2004).
20. T. Schreiber, *Measuring information transfer*, Phys. Rev. Lett. **85**, 461 (2000).
21. M. Paluš and A. Stefanovska, *Direction of coupling from phases of interacting oscillators: An information-theoretic approach*, Phys. Rev. E **67**, 055201(R) (2003).
22. J. Jamsek and A. Stefanovska, P.V.E. McClintock, *Nonlinear cardio-respiratory interactions revealed by time-phase bispectral analysis*, Phys. Med. Bio. **49**, 4407–4425 (2004).
23. M. Richter, T. Schreiber, and D.T. Kaplan, *Fetal ECG extraction with nonlinear state-space projections*, IEEE Eng. Med. Biol. Mag. **45**, 133–137 (1998).
24. R. Hegger, H. Kantz, and T. Schreiber, *Practical implementation of nonlinear time series methods: The TISEAN package*, Chaos **9**, 413–435 (1999).
25. H. Kantz and T. Schreiber, *Human EGG: Nonlinear deterministic versus stochastic aspects*, IEE Proceedings – Science Measurement and Technology **145**, 279–284 (1998).
26. M.-C. Wu and C.-K. Hu, *Empirical mode decomposition and synchrogram approach to cardiorespiratory synchronization*, Phys. Rev. E **73**, 051917 (2006).
27. D. Gabor, *Theory of communication*, J. Inst. Electron Eng. **93**, 429–457 (1946).
28. S. Blanco, R. Q. Quiroga, O.A. Rosso, and S. Kochen, *Time-frequency analysis of electroencephalogram series*, Phys. Rev. E **51**, 2624 (1995).
29. S. Blanco, C.E. D'Attellis, S.I. Isaacson, O.A. Rosso, and R.O. Sirne, *Time-frequency analysis of electroencephalogram series. II. Gabor and wavelet transforms*, Phys. Rev. E **54**, 6661 (1996).
30. S. Blanco, A. Figliola, R.Q. Quiroga, O.A. Rosso, and E. Serrano, *Time-frequency analysis of electroencephalogram series. III. Wavelet packets*, Phys. Rev. E **57**, 932 (1998).
31. K. Ohashi, L.A.H. Amaral, B.H. Natelson, and Y. Yamamoto, *Asymmetrical singularities in real-world signals*, Phys. Rev. E **68**, 065204(R) (2003).
32. K. Karhunen, *Uber lineare methoden in der wahrscheinlichkeits-rechnung*, Ann. Acad. Sci. Fennicae, ser. A1, Math. Phys. **37** (1946).
33. M.M. Loéve, *Probability theory*, Princeton, NJ, Van Nostrand (1955).
34. N.E. Huang, Z. Shen, S.R. Long, M.C. Wu, H.H. Shih, Q. Zheng, N.-C. Yen, C.-C. Tung, and H.H. Liu, *The empirical mode decomposition and the Hilbert spectrum for nonlinear and non-stationary time series analysis*, Proc. R. Soc. Lond. A **454**, 903–995 (1998).
35. N. Iyengar, C.-K. Peng, R. Morin, A. L. Goldberger, and L. A. Lipsitz, *Age-related alterations in the fractal scaling of cardiac interbeat interval dynamics*, Am. J. Physiol. **271**, 1078–1084 (1996). Data sets are available from http://physionet.org/physiobank/database/fantasia/
36. N.E. Huang, M.C. Wu, S.R. Long, S.S. P. Shen, W. Qu, P. Gloersen, and K.L. Fan, *A confidence limit for the empirical mode decomposition and Hilbert spectral analysis*, Proc. R. Soc. Lond. A **459**, 2317–2345 (2003).
37. M. G. Rosenblum and A. S. Pikovsky, *Detecting direction of coupling in interacting oscillators*, Phys. Rev. E **64**, 045202(R) (2001).
38. M.-C. Wu, M.-C. Huang, Y.-C. Yu, and T. C. Chiang, *Phase distribution and phase correlation of financial time series*, Phys. Rev. E **73**, 016118 (2006).
39. M.-C. Wu, *Phase correlation of foreign exchange time series*, Physica A **375**, 633–642 (2007).
40. M.-C. Wu, *Phase statistics approach to time series analysis*, J. Korean Phys. Soc. **50**, 304–312 (2007).

Part IV
Cognitive and Neurosciences

Brain Dynamics and Modeling in Epilepsy: Prediction and Control Studies

Leonidas Iasemidis, Shivkumar Sabesan, Niranjan Chakravarthy, Awadhesh Prasad and Kostas Tsakalis

Abstract Epilepsy is a major neurological disorder characterized by intermittent paroxysmal neuronal electrical activity, that may remain localized or spread, and severely disrupt the brain's normal operation. Epileptic seizures are typical manifestations of such pathology. It is in the last 20 years that prediction and control of epileptic seizures has been the subject of intensive interdisciplinary research. In this communication, we investigate epilepsy from the point of view of pathology of the dynamics of the electrical activity of the brain. In this framework, we revisit two critical aspects of the dynamics of epileptic seizures – the seizure predictability and seizure resetting – that may prove to be the keys for improved seizure prediction and seizure control schemes. We use human EEG data and the concepts of spatial synchronization of chaos, phase and energy to first show that seizures could be predictable in the order of tens of minutes prior to their onset. We then present additional statistical evidence that the pathology of the brain dynamics prior to seizures is reset mostly upon seizures' occurrence, a phenomenon we have called seizure resetting. Finally, using a biologically-plausible neural population mathematical model that can exhibit seizure-like behavior, we provide evidence for the effectiveness of a recently devised seizure control scheme we have called "feedback decoupling". This scheme also provides an interesting dynamical model for ictogenesis (generation of seizures).

1 Introduction

Epileptic seizures are manifestations of epilepsy, a neurological dynamical disorder second only to stroke. Of the world's $\simeq$50 million people with epilepsy, about 1/3 has seizures that are not controlled by anti-convulsant medication. One of the most disabling aspects of epilepsy is the seemingly unpredictable nature of seizures. If seizures cannot be controlled, the patient experiences major limitations in family, social, educational, and vocational activities. These limitations have profound

L. Iasemidis (✉)
The Harrington Department of Bioengineering and Electrical Engineering,
Arizona State University, Tempe, Arizona, USA

effects on the patient's quality of life, as well as on his or her family [1–3]. In addition, *status epilepticus*, a life-threatening condition where seizures occur continuously, is treated only upon extreme intervention [4]. Until recently, the general belief in the medical community was that epileptic seizures could not be anticipated. Seizures were assumed to be abrupt transitions that occurred randomly over time. However, theories based on reports from clinical practice and scientific intuition, like the "reservoir theory" postulated by Lennox [5], existed and pointed out to the direction of seizure predictability. Various feelings of auras, that is, patients' reports of sensations of an upcoming seizure, also exist in the medical literature. Penfield [6] was the first to note changes in the cerebral blood flow prior to seizures. Deterministically predictable occurrences of seizures (reflex seizures) in a small minority (about 3 to 5%) of epileptic patients have been reported as a result of various sensory stimuli [7, 8]. These theories and facts have implied that seizures might be predictable.

The ability to predict epileptic seizures well prior to their occurrences may lead to novel diagnostic tools and treatment of epilepsy. Evaluation of anti-epileptic drugs and protocols, in terms of duration of patients' seizure susceptibility periods and/or preictal (before a seizure) periods detected by seizure prediction algorithms, may lead to the design of new, more effective and with less side effects drugs for early disruption of the epileptic brain's route towards a seizure. Electromagnetic stimulation and/or administration of anti-epileptic drugs (AEDs) at the beginning of the preictal period may disrupt the observed dynamical entrainment of normal brain sites with the epileptogenic focus (the area that first exhibits the electrographic onset of ictal activity), and lead to a significant reduction of epileptic seizures. Aside from their immediate clinical applications to epilepsy, successful seizure prediction and control algorithms could be useful for investigations into a wide variety of other complex, nonstationary and spatio-temporal biological and physical systems that undergo intermittent transitions.

The 80s saw the emergence of new signal processing methodologies, based on the mathematical theory of nonlinear dynamics, for the study of spontaneous formation of organized spatial, temporal or spatiotemporal patterns in physical, chemical and biological systems [9, 10, 12, 13]. These methodologies quantify the signal structure from the perspective of dynamical invariants (e.g. dimensionality of the attractor through correlation dimension, or divergence of trajectories through the largest Lyapunov exponent), and are a drastic departure from the signal processing techniques based on the linear model (e.g. Fourier analysis). In 1988, a small group at the University of Michigan, Ann Arbor, led by Iasemidis, Sackellares and Williams, reported the first application of nonlinear dynamics to clinical epilepsy [14]. That also led to the first NIH (National Institute of Health) supported clinical investigation into the nonlinear dynamics of epileptic seizures ("Dynamical studies in temporal lobe epilepsy", NIH-NINCDS RO1 NS31451). This group started to analyze continuous, multichannel, preictal (before seizure), ictal (during seizure) and postictal (after seizure) EEG from epileptic patients with temporal lobe epilepsy devising new and modifying existing measures from the theory of chaos to quantify the rate of divergence of trajectories (Lyapunov exponent) for the analysis of

EEG in epilepsy. The central concept was that seizures represented transitions of the epileptic brain from its "normal" less ordered (chaotic) state to an abnormal, more ordered state, and back to a "normal" state along the lines of chaos-to-order-to-chaos transitions [15–24].

The dynamical modeling hypothesis changed some long-held beliefs about seizures. Iasemidis et al. reported the first evidence that the transition to epileptic seizures may be consistent with a deterministic process [15, 20], and that ictal electroencephalogram (EEG) can be better modeled as an output of a nonlinear than a linear system [17]. The existence of long-term preictal periods (order of minutes) was shown using nonlinear dynamical analysis of subdural arrays [16], and raised the feasibility of seizure prediction algorithms by monitoring the temporal evolution of the short-term Lyapunov exponents (STL_{max}) [22–24]. The possibility of focus localization and seizure detection was also reported with the same technique in 1990 and 1994 respectively [18–21]. Since these initial results, several research groups in the world started to work in the area of seizure prediction (see [25] for a review). Elger and Lehnertz investigated the spatio-temporal dynamics of the epileptic focus in 1994 [26, 27], while Scott and Schiff directed attention to the time structure of inter-ictal spikes [28]. Lopes da Silva et al. who have been developing neurophysiology-driven dynamical models for EEG activity since the late 70s, quantified state bifurcations in epileptogenesis [29]. Iasemidis and Sackellares perfected their STL_{max} technique with the use of optimization techniques and the critical mass hypothesis to predict seizures [30–36]. The fundamental issue that surfaced through this group's investigations in seizure predictions was the importance to locate and use only the channels that carry information about an impending seizure. Changes in the spatio-temporal structure of the EEG can in principle also be quantified by sophisticated linear methods involving wavelet decompositions, coherence, and pattern recognition methods such as artificial neural networks, and fuzzy approaches. Several research groups are employing this approach towards the detection of the preictal period [37, 38]. However, no prospective results on seizure prediction (i.e. seizure prediction from long-term continuous EEG data using information only from past seizure occurrences) have been reported in the literature with any of the above measures yet, with the exception of Iasemidis et al.'s framework of analysis. The important conclusion from all these different techniques is an accumulation of evidence that there are measurable differences in the EEG prior to seizure onset that can be utilized for epileptic seizure prediction.

Iasemidis's group has reported a progressive preictal increase of spatiotemporal entrainment/synchronization between critical sites of the brain as the precursor of epileptic seizures. The algorithm used was based on the convergence of short-term maximum Lyapunov exponents (STL_{max}) among critical electrode sites selected adaptively. This observation has been successfully implemented in the prospective prediction of epileptic seizures [39]. Global optimization techniques were applied for selecting the critical groups of electrode sites to observe preictal entrainment. Seizure anticipation times of about 71.7 min with a false prediction rate of 0.12 per h were reported. To further relate these findings to the mechanism of epileptogenesis, these investigators found that majority of seizures in patients with temporal lobe epilepsy (TLE) irreversibly reset (disentrain) postictally the observed preictal dynamical entrainment [40, 41]. This supports the hypothesis that seizures do not occur as long as there is no need to reset the brain. Last, but not least, this group of researchers have also shown through simulations that, in chaos-to-order-to-chaos transitions of general models of spatially coupled chaotic oscillators, with increase/decrease of coupling convergence/divergence of the oscillators' STL_{max} resembles the observed preictal entrainment and postictal dynamical disentrainment of the STL_{max} of critical brain sites. In addition, the model exhibited hysteresis, a phenomenon that is also observed in the epileptic transition into and out of seizures [41]. This dynamical view leads to a characterization of the seizure itself as a mechanism that the brain has developed to reset the preictal entrainment when a critical mass of sites, or a mass of specific, critical sites, is recruited.

In order to provide insights into the development of feedback control strategies for suppression of seizures in the epileptic brain, and motivated by analysis and results of burst phenomena in adaptive systems [42, 43], Tsakalis and Iasemidis [44] postulated the existence of an internal pathological feedback action in the epileptic brain that lacks the ability to compensate for excessive increases in the network coupling. This situation eventually leads to seizure-like transitions [44–47]. A precursor of this scenario is an abnormal increase in synchronization that cannot be regulated quickly enough by the existing pathological internal feedback mechanism. Using a control-oriented approach, a functional model of an external feedback stimulation paradigm was developed. During periods of abnormally high synchrony, this scheme provides appropriate "desynchronizing feedback" to maintain "normal" synchronization levels between neural populations. This feedback control view of epileptic seizures and the developed seizure control strategies have been validated using coupled chaotic oscillator models as well as biologically plausible neurophysiologic models [46, 47].

In summary, from our group's past and ongoing research, the following three central results about epileptic seizures have emerged. First, we have shown that seizures are manifestations of recruitment of brain sites in an abnormal hypersynchronization. The onset of such recruitment occurs long before a seizure and progressively culminates into a seizure. Seizures appear to be bifurcations of a neural network that involve a progressive coupling of the focus with the normal brain sites during a preictal period that may last days to tens of minutes. Thus, identification of such a preictal period may constitute the basis for predicting an impending seizure well in advance.

Second, postictally, time-irreversible resetting of the observed preictal dynamical recruitment occurs (via a hysteretic loop). Preictal and postictal periods could be mathematically defined and detected from the EEG. Complete or partial resetting of the preictal entrainment of the epileptic brain after a seizure may affect the route to the subsequent seizure. This may contribute to the observed nonstationary nature of the seizure occurrences. Therefore, it is expected that estimation of the magnitude of resetting at a seizure will improve our understanding of the brain's route to subsequent seizures, and may even lead to better seizure prediction and control.

Brain Dynamics and Modeling in Epilepsy: Prediction and Control Studies 189

Third, through control-oriented modeling, a feedback control view of epileptic seizures has been postulated, wherein epileptic seizures are hypothesized to be a result of the inability of the internal feedback/regulatory mechanisms of the brain to track excessive synchronization changes between the epileptogenic focus and other brain areas prior to a seizure.

We herein present results on the preictal entrainment and brain resetting in EEG data recorded from epileptic patients, as well as on the generation and control of seizure-like behavior in a biologically-plausible mathematical model. Accordingly, the first goal was to identify the most reliable synchronization measures, across seizures in the same patient and across patients, that also issue the earliest warnings of upcoming seizures. Second, in order to further shed light on the mechanisms of seizures occurrence, the concept of seizure resetting is revisited. Third, a feedback control scheme for generation and suppression of epileptic seizures is shown, after we suitably modify a coupled neural population model from the literature to exhibit "seizures".

In the next Section 2, we first describe the available EEG data, and then the results of the application of the different synchronization measures to the predictability of the recorded epileptic seizures. In Section 3, we elaborate on the idea of brain resetting. Novel measures that detect resetting, as well as the sensitivity and specificity of resetting at seizure points, are investigated. We describe a feedback, systems-based modeling of the "preictal" dynamics via simulations on coupled neural populations in Section 4. In Section 5, we present a closed-loop seizure control strategy for the epilepsy-prone model in Section 4. We further discuss these results and present our conclusions in Section 6.

2 Predictability of Epileptic Seizures

In the present section, three of the most frequently utilized measures of dynamical synchronization/entrainment, namely classical energy (E), phase (Φ) and short-term Lyapunov exponent (STL_{max}), are compared on the basis of their ability to detect preictal changes. It is noteworthy that these three quantities, in the case of the complex exponential basis signal $x(t) = Ae^{\alpha t}e^{j(\omega t + \phi)}$, correspond to $Ae^{\alpha t}$, $\omega t + \phi$, and α respectively, where α is the real part of the pole of the Laplace transform of x(t) and is equal to STL_{max}, phase $\Phi = \omega t + \phi$ is the imaginary part of the pole and it depends on ω, and energy E depends on A and α. Due to the current interest in the field, and the proposed measures of energy and phase as alternatives to STL_{max} for seizure prediction [38, 48, 49], it is important to comparatively evaluate each of the three measures' seizure predictability (anticipation) capabilities.

Quadratic integer programming techniques of global optimization were applied to select the optimal electrode sites per measure and seizure, that is the ones that exhibit maximum synchronization for every recorded seizure. Results from such an analysis of 43 seizures, recorded from two patients with temporal lobe epilepsy, showed that optimal sites selected on the basis of STL_{max} 10 min before a seizure appear to have longer and more consistent preictal trends prior to the majority of seizures than the ones exhibited by the optimal sites selected within the same period on the basis of the other two measures of synchronization. This section is organized as follows: First, we describe the EEG data analyzed in this manuscript. We then summarize the estimation of the three measures of synchronization from EEG. Statistical yardsticks used to quantify the performance of each measure in detecting preictal dynamics are provided next. Subsequently, the formulation of the quadratic integer programming problem for the selection of critical electrode sites from a measure of synchronization for seizure predictability is discussed. Finally, seizure predictability results (e.g. seizure predictability times) from the application of these methods to EEG are presented.

2.1 Description of EEG Data

All 43 seizures recorded from two epileptic patients with temporal lobe epilepsy at the epilepsy monitoring unit (EMU) (see Table 1) were analyzed by the methodologies described below. The EEG signals were recorded from 6 different areas of the brain by 28 electrodes (see Fig. 1 for the electrode montage). Typically, periods from 3 h before and 1 h after each seizure were analyzed in search of dynamical synchronization and for estimation of seizure predictability times.

The patients in the study underwent a stereotactic placement of bilateral depth electrodes (RTD1 to RTD6 in the right hippocampus, with RTD1 adjacent to right amygdala; LTD1 to LTD6 in the left hippocampus with the LTD1 adjacent to the left amygdala; the rest of the LTD, RTD electrodes extending posterior through the hippocampi). Two subdural strip electrodes were placed bilaterally over the orbitofrontal lobes (LOF1 to LOF4 in the left and ROF1 to ROF4 in the right lobe, with LOF1, ROF1 being most mesial and LOF4, ROF4 most lateral). Two subdural strip electrodes were placed bilaterally over the temporal lobes (LST1 to LST4 in the left and RST1 to RST4 in the right, with LST1, RST1 being more mesial and LST4, RST4 being more lateral).

Video/EEG monitoring was performed using the Nicolet BMSI 4000 EEG machine. EEG signals were recorded using an average common reference with band pass filter settings of 0.1 Hz–70 Hz. The data were sampled at 200 Hz with a 10-bit quantization and recorded on VHS tapes continuously over days via 3 time-

Table 1 Patient and EEG data characteristics

Patient ID	Number of electrode sites	Location of epileptogenic focus	Seizure type	Duration of EEG (days)	Number of recorded seizures	Number of analyzed seizures	Analyzed EEG per seizure (hours)
1	28	RTD	C	9.06	24	24	4
2	28	RTD	C, SC	6.07	19	19	4

Seizure types: C (clinical); SC (subclinical)

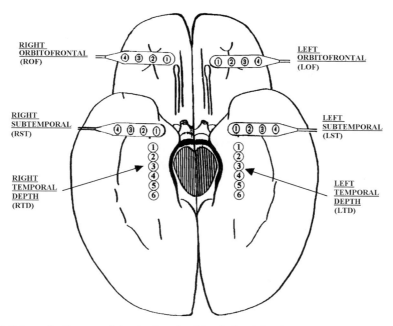

Fig. 1 Schematic diagram of the depth and subdural electrode placement. This view from the inferior aspect of the brain shows the approximate location of depth electrodes, oriented along the anterior-posterior plane in the hippocampi (RTD – right temporal depth, LTD – left temporal depth), and of subdural electrodes located beneath the orbitofrontal and subtemporal cortical surfaces (ROF – right orbitofrontal, LOF – left orbitofrontal, RST – right subtemporal, LST – left subtemporal)

interleaved VCRs. Decoding of the data from the tapes and transfer to computer media (hard disks, DVDs, CD-ROMs) was subsequently performed off-line. The seizure predictability analysis also was performed retrospectively (off-line).

2.2 Measures of EEG Synchronization

2.2.1 Energy Profiles *(E)*

The classical energy of a signal x(t) over time is calculated as the sum of its magnitude squared over a time period T, like $E(t) = \sum_{\varepsilon=t}^{t+T} |x(\varepsilon)|^2$. For EEG analysis, E values are calculated over consecutive non-overlapping segments of data, each segment being T second in duration (here we use $T = 10.24$ sec), from different locations in the brain over time t. Examples of E profiles over time from two electrode sites that show synchronization before a seizure are given in the left panel of Fig. 2a below. The highest E values were observed during the ictal period. This

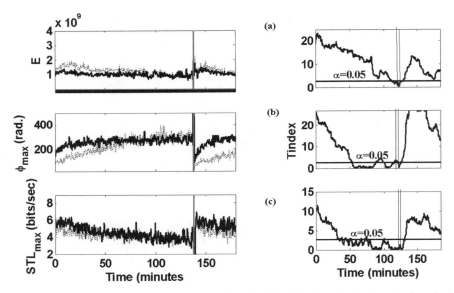

Fig. 2 Long-term synchronization prior to a seizure (Patient 1; seizure 12). **Left Panels**: (*a*) *E* profiles over time of two electrode sites (RST1, ROF2) selected to be mostly synchronized 10 min prior to the seizure. (*b*) $\Phi_{\max}$ profiles of two electrode sites (RTD1, RST2) selected to be mostly synchronized 10 min prior to the seizure. (*c*) $STL_{\max}$ profiles of two electrode sites (RTD2, LOF3) selected to be mostly synchronized 10 min prior to the seizure. **Right Panels**: Corresponding T-index curves for the sites and measures depicted in the left panels. Vertical lines illustrate the period over which the effect of the ictal period is present in the estimation of the T-index values, since 10-min windows $w_1(t)$ move forward in time every 10.24 sec over the values of the measure profiles in the left panels. Seizure lasted for 2 min, hence the depicted period between the vertical lines is 12 min

pattern roughly corresponds to the typical observation of higher amplitudes in the original EEG signal ictally.

2.2.2 Maximum Phase Profiles ($\Phi_{\max}$)

The notion of phase synchronization was introduced by Huygens [50] in the 17th century for two coupled frictionless harmonic oscillators oscillating at different angular frequencies ω_1 and ω_2 such that $\frac{\omega_1}{\omega_2} = \frac{m}{n}$. In this classical case, phase synchronization is usually defined as locking of the phases of the two oscillators in terms of $\varphi_{n,m} = n\phi_1(t) - m\phi_2(t) =$ constant, where *n* and *m* are integers, ϕ_1 and ϕ_2 denote the phases of the oscillators, and $\varphi_{n,m}$ is defined as their constant *relative phase*. In order to investigate synchronization in chaotic systems, Rosenblum et al. [51] relaxed this condition of *phase locking* by a weaker condition of *phase synchronization* (since $\frac{\omega_1}{\omega_2}$ may be an irrational real number and each system may contain power at other frequencies besides a dominant frequency):

$$|\varphi_{n,m}| = |n\phi_1(t) - m\phi_2(t)| < \text{constant}. \tag{1}$$

The estimation of instantaneous phases $\phi_1(t)$ and $\phi_2(t)$ is nontrivial for many nonlinear model systems, and even more difficult when dealing with noisy time series of unknown characteristics.

Different approaches have been proposed in the literature for the estimation of instantaneous phase of a signal. In the analysis that follows, we take the *analytic signal* approach for phase estimation [52, 53] that defines the *instantaneous phase* of an arbitrary signal $s(t)$ as:

$$\phi(t) = \arctan \frac{\tilde{x}(t)}{x(t)},$$

where

$$\tilde{x}(t) = \frac{1}{\pi} P.V. \int_{-\infty}^{+\infty} \frac{x(\tau)}{t - \tau} d\tau$$

is the *Hilbert Transform* of the signal $x(t)$ (*P.V.* denotes the Cauchy Principal Value). The Hilbert transformation is equivalent to a special kind of filtering of $x(t)$ in which amplitudes of the spectral components are left unchanged, while their phases are altered by $\pi/2$, positively or negatively according to the sign of ω. For details of the practical estimation of $\phi(t)$ via the Hilbert transform, the tapering of the data by a Hamming window and subsequent utilization of the Fourier Transform [54]. The $\phi(t)$ from EEG data were estimated per non-overlapping moving windows of 10.24 sec in duration and per electrode site. An array of $\phi(t)$ values are returned for each window, equal in number to the number of EEG data points ($x(t)$) contained in this window. Then, the maximum value (Φ_{max}) of the phase values per window is estimated and used in subsequent analysis. An example of Φ_{max} profiles at two electrode sites over time is given in Fig. 2b (left panel). The preictal, ictal and postictal states corresponded to medium, high and lower values of Φ_{max} respectively. The highest Φ_{max} values were observed during the ictal period. Higher Φ_{max} values were observed during the preictal period than the postictal period. This pattern roughly corresponds to the typical observation of higher frequencies in the original EEG signal ictally, and lower EEG frequencies postictally.

2.2.3 Chaos profiles (*STL*)

Under certain conditions, through the method of delays described by Packard et al. [55] and Takens [56], sampling of a single variable of a system over time could determine all state variables of the system that are related to an observed state of the system. In the case of the EEG, this method can be used to reconstruct a multi-dimensional state space of the brain's electrical activity from a single EEG channel recording at a corresponding brain site. Thus, in such an embedding, each state in the state space is represented by a vector $X(t)$, whose components are the delayed versions of the original single-channel EEG time series $x(t)$, that is:

$$X(t) = \left(x(t), x(t+\tau), \ldots, x(t+(d-1)\cdot\tau)\right) \tag{2}$$

where τ is the time delay between successive components of $X(t)$, and d is a positive integer denoting the embedding dimension of the reconstructed state space. Plotting $X(t)$ in the thus created state space produces the state portrait of a subsystem (brain site) of a spatially distributed system (brain) where $x(t)$ is recorded from. A steady state of such a subsystem is chaotic if at least the maximum L_{max} of its Lyapunov exponents (L_s) is positive [57, 58].

Of the many different methods used to estimate d of an object in the state space, each has its own practical problems [59]. The measure most often used to estimate d is the state space correlation dimension ν, where $d \geq 2\nu + 1$. Methods for calculating ν from experimental data have been described in [60] and were employed in our work to approximate ν in the ictal state. The brain, being nonstationary, is never in a steady state in the strictly dynamical sense at any location. Arguably, activity at brain sites is constantly moving through approximately steady states, which are functions of certain parameter values at a given time. According to bifurcation theory [61], when these parameters change slowly over time, or the system is close to a bifurcation, dynamics slow down and conditions of stationarity are better satisfied. In the ictal state, temporally ordered and spatially synchronized oscillations in the EEG usually persist for a relatively long period of time (in the range of minutes). Dividing the ictal EEG into short segments ranging from 10.24 sec to 50 sec in duration, estimation of ν from ictal EEG has given values between 2 and 3 [31, 32], implying the existence of a low-dimensional manifold in the ictal state, which we have called "epileptic attractor". Therefore, we used an embedding dimension $d = 7$ to properly reconstruct this epileptic attractor.

Although d of interictal (between seizures) "steady state" EEG data is expected to be higher than that of the ictal state, a constant embedding dimension $d = 7$ has been used to reconstruct all relevant state spaces over the ictal and interictal periods at different brain locations. The advantages of using such a small embedding dimension are that: (a) irrelevant information to the epileptic transition in dimensions higher than 7 would not influence much the values of the dynamical measures, and (b) estimation of the dynamical measures suffers less from the small number of data points allowable per moving window (short windows are used to address the nonstationarity of the EEG). The disadvantage is that critical information for the transition to seizures that may exist in dimensions higher than 7 would not be captured.

The Lyapunov exponents measure the average information flow (bits/sec) a system produces along local eigenvectors of its movement in its state space. Methods for calculating these dynamical measures from experimental data have been published [16, 31]. The estimation of the largest Lyapunov exponent (L_{max}) in a chaotic system has been shown to be more reliable and reproducible than the estimation of the remaining exponents [62], especially when the value of d is not known and changes over time, as it is the case in high-dimensional and nonstationary data like the interictal EEG. A measure developed to estimate an approximation of L_{max} from nonstationary data was called STL_{max} (Short-Term maximum Lyapunov exponent) [16, 31]. The STL_{max} is estimated from sequential EEG epochs of 10.24 sec, recorded from electrodes in multiple brain sites, to create a set of STL_{max} profiles over time (resulting to one STL_{max} value per epoch, one STL_{max} profile per recording site) that characterizes the spatio-temporal chaotic signature of the epileptic brain. The STL_{max} profiles at two electrode sites are shown in Fig. 2c (left panel). These figures show the evolution of STL_{max} as the brain progresses from interictal to ictal to postictal states. The seizure onset is characterized by a sudden drop in STL_{max} values, with a consequent steep rise in STL_{max} and higher values in the postictal than the ictal period, denoting a chaos-to-order-to-chaos transition. What is also observed is a convergence of the STL_{max} profiles long before the seizure's onset. We have called this convergence "dynamical entrainment", because it is progressive (entrainment) and involves measures of the dynamics (dynamical) of the underlying subsystems in the brain. This has constituted the basis for the development of the first prospective epileptic seizure prediction algorithms [39,63–65]. For the purpose of this communication, we will refer to this phenomenon as dynamical synchronization or synchronization of the brain dynamics.

2.2.4 T-Index as a Measure for Synchronization of EEG Dynamics

A statistical measure of synchronization between two electrodes i and j, with respect to a measure (e.g. STL_{max}, E or Φ_{max}) of their dynamics, has been developed before (e.g. [31]) and is described below. Specifically, the T_{ij} between measures at electrode sites i and j and time t is defined as:

$$T_{ij}(t) = \frac{\hat{D}_{ij}(t)}{\hat{\sigma}_{ij}(t)/\sqrt{m}},$$

where $\hat{D}_{ij}(t)$ and $\hat{\sigma}_{ij}(t)$ denote the sample mean and standard deviation respectively of all the m differences $D_{ij}(t)$ between a measure's values (one value per 10.24 sec) at electrodes i and j, within a moving window $w_1(t) = [t, t - m^*10.24 \sec]$. If the true mean $\mu_{ij}(t)$ of the differences $D_{ij}(t)$ is equal to zero, and $\hat{\sigma}_{ij}(t)$ are independent and normally distributed, $T_{ij}(t)$ is asymptotically distributed as the t-distribution with $(m-1)$ degrees of freedom. We have shown that these independence and normality conditions are satisfied in EEG [63]. Therefore, we define desynchronization between electrode sites i and j when $\mu_{ij}(t)$ is significantly different from zero at a significance level α. The desynchronization condition between the electrode sites i and j, as detected by the paired t-test, is

$$T_{ij}(t) > t_{\alpha/2, m-1} = T_{th} \qquad (3)$$

where $t_{\alpha/2, m-1}$ is the $100 \cdot (1-\alpha/2)\%$ critical value of the t-distribution with $m-1$ degrees of freedom. If $T_{ij}(t) \leq t_{\alpha/2, m-1}$ (which means that we do not have satisfactory statistical evidence at the α level that the differences of values of a measure between electrode sites i and j within the time window w(t) are not zero), we consider that sites i and j are synchronized with each other (with respect to the utilized measure of synchronization) at time t. Using $\alpha = 0.01$ and $m = 60$, the threshold $T_{th} = 2.662$. It is noteworthy that similar STL_{max}, E, or Φ_{max} values (i.e. when these measures are synchronized) at two electrode sites do not necessarily mean that these sites also interact. However, when there is a progressive convergence (synchronization) over time of the measures at these sites, the probability that they are unrelated diminishes. This is exactly what occurs before seizures, and it is illustrated in the right panels of Fig. 2 for all three measures considered herein. A progressive synchronization in all measures, as quantified by $T_{ij}(t)$, is observed preictally. Note that, for each measure, different critical sites are synchronized. These critical sites per measure were selected according to the procedure described below in Section 2.3.

2.3 Selection of Optimal Sites per Measure via Global Optimization

Not all brain sites are progressively synchronized prior to a seizure. The selection of the ones that do (critical sites) can be formulated as a global optimization problem that seeks to minimize the distance between the dynamical measures at these sites within a time window $w_1(t)$. Motivated by the application of the Ising model to phase transitions, we have adapted quadratic bivalent (zero-one) programming techniques to optimally select the critical electrode sites [64, 66] that, during the preictal transition, minimize the objective function of the distance between values of measures of synchronization in all pairs of these brain sites. This procedure corresponds to minimization of the distances of each STL_{max}, Φ_{max}, E measure in pairs of brain sites.

More specifically, we considered the integer bivalent 0–1 problem:

$$\min x^t T x \text{ with } x \in \{0,1\}^n \text{ subject to the constraint } \sum_{i=1}^{n} x_i = k \tag{4}$$

where n is the total number of available electrode sites, k is the number of sites to be selected, and x_i are the (zero / one) elements of the n-dimensional vector x. The elements of the T matrix T_{ij}, $i = 1, \ldots n$ and $j = 1, \ldots n$ for each measure are previously defined in Section 2.2.4. If the above constraint is included in the objective function $x^t T x$ by introducing the penalty

$$\mu = \sum_{j=1}^{n} \sum_{i=1}^{n} |T_{ij}| + 1, \tag{5}$$

the optimization problem becomes equivalent to an unconstrained global optimization problem

$$\min \left[x^t T x + \mu \left(\sum_{i=1}^{n} x_i - k \right)^2 \right], \text{ where } x \in \{0,1\}^n \tag{6}$$

The electrode site i is selected if the corresponding element x_i^* in the n-dimensional solution x^* of (6) is equal to 1. In our comparison of the three measures of dynamical synchronization herein we have chosen the value of $k = 5$ [34, 63]. The optimization procedure for the selection of critical sites was carried out within the preictal window $w_1(t^*) = [t^*, t^*-10\,\text{min}]$ over a measure's profiles, where t^* is the time at a seizure's onset. The average T-index across all the T-indices of all possible pairs of the selected sites is then generated and followed backward in time from each seizure's onset t^*. In the following sections, for simplicity, we denote these spatially averaged T-index values by "T-index". For demonstration purposes, the corresponding T-indices between two critical sites ($k = 2$) for each measure is depicted in Fig. 2 (right panel) before and after one of the recording seizures in patient 1. After the optimal sites selection from each of the three measures E, STL_{max} and Φ_{max} profiles within the preictal window $w_1(t^*)$, the average T-index of the selected sites was estimated backward and forward in time with respect to t^*.

2.4 Estimation of Seizure Predictability Time

The predictability time T_p for a given seizure is defined as the period before a seizure's onset during which synchronization between critical sites remains highly statistically significant (i.e. T-index < 2.662). Each measure of synchronization gives a different T_p for a seizure. In the estimation of T_p, to compensate for possible random oscillations of the T-index profile, we average T_p values within windows $w_2(t)$ moving on the T-index profile backward in time from a seizure's onset. We used the same length of $w_2(t)$ as the one of $w_1(t)$ so that T_{th} is the same for both windows for $\alpha = 0.01$. Then, T_p is estimated by the following procedure: The time average $\overline{T}$ -index of the T-indices over 3 h before a seizure, within $w_2(t) = [t, t-10\,\text{min}]$, where t* > t > t*–170 min, is continuously estimated until $\overline{T}$ -index is less than or equal to T_{th}.. When $t = t_0$: $\overline{T}$ -index > T_{th}, the T_p is defined as $T_p = t^*-t_0$. This predictability time estimation is portrayed in Fig. 3 (where $t^* = 0$). The longer the T_p, the longer synchronization is observed prior to a seizure. The comparison of the seizure predictability times T_p, estimated by each of the three measures STL_{max}, E, Φ_{max}, is presented in the next section.

2.5 Predictability Results

Per each of the three synchronization measures, the 5 most synchronized sites within 10 min prior to each of the 43 recorded seizure onset were selected by the optimization procedure described in the previous Section 2.3. The spatially averaged T-indices across these sites were estimated per seizure over time (3 h before to 1 h after each seizure). Then, the predictability times T_p were estimated for each seizure and dynamical measure. The algorithm for estimation of T_p delivered

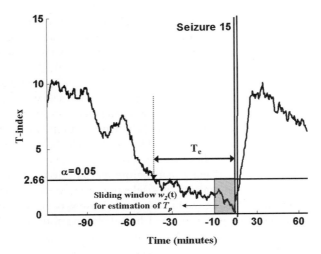

Fig. 3 Estimation of the seizure predictability time T_p from the T-index profile of the selected critical sites for a seizure (seizure 15, Patient 1). The average $\overline{T}$-index of the critical sites within a moving window $w_2(t)$ is continuously estimated as the window $w_2(t)$ moves backward in time from the seizure onset. When the average $\overline{T}$-index $> T_{th} = 2.662$, T_p is conservatively set equal to the right endpoint of $w_2(t)$

visually agreeable predictability times for all T-index profiles. The T-index profiles preictally decreased towards a seizure in a near-monotonic fashion (see Fig. 2).

2.5.1 Comparative performance of Energy, Phase, STL_{max} measures

The average predictability time obtained for the measure of STL_{max} across all seizures was 61.6 and 71.69 min for patients 1 and 2 respectively (see Table 2 – underlined values). The measure of classical energy E, used under a univariate

Table 2 Statistics of seizure predictability times T_p across seizures per patient and synchronization measure. [$T_{p-\text{opt}}$ from five optimal sites (underlined) versus average $T_{p-\text{random}}$ from 10 groups of five randomly selected sites (not underlined)]

T_p(minutes)							
Measure	Patient 1 (24 seizures)			Patient 2 (19 seizures)		
	Mean	Std.	$P(T_{pt} \leq T_p)$	Mean	Std.	$P(T_{pt} \leq T_m)$
STL_{max}	*61.60*	*45.50*	$P < 0.0005$	*71.69*	*33.62*	$P < 0.0005$
	5.80	10.32		11.29	13.54	
E	*13.72*	*11.50*	$P < 0.002$	*27.88*	*26.97*	$P < 0.004$
	5.32	7.40		7.44	9.27	
Φ_{max}	*39.09*	*20.88*	$P < 0.0005$	*47.33*	*33.34*	$P < 0.0005$
	7.19	6.55		7.37	9.12	

analysis (estimation per electrode site without subsequently considering bivariate synchronization measures), has been shown to lack consistent predictive ability for a seizure [67]. Furthermore, its predictive performance was shown to deteriorate from postictal changes of preceding seizures and changes during sleep-wake cycles [48]. By our spatiotemporal (multivariate) synchronization analysis of E profiles, we found average predictability times of 13.72 and 27.88 min for patients 1 and 2 respectively, a significant improvement in their performance over what reported in the literature before. For the measure of phase synchronization Φ_{max}, the average predictability time values were 39.09 and 47.33 min for patients 1 and 2 respectively.

Dynamical synchronization analysis using STL_{max} consistently resulted to longer predictability times T_p than the ones by the other two measures (see Table 2 – underlined values). Between the other two measures, the phase synchronization measure outperformed the linear, energy-based measure and, for some seizures, it even had comparable performance to that of STL_{max} – based synchronization. These results are consistent with the synchronization observed in coupled non-identical chaotic oscillator models: an increase in coupling between two oscillators initiates generalized synchronization (best detected by STL_{max}), followed by phase synchronization (detected by phase measures), and upon further increase in coupling, amplitude synchronization (detected by energy measures) [68–73].

2.5.2 Improved Predictability via Global Optimization

We tested the hypothesis that global optimization for the selection of critical sites per seizure is crucial in order to achieve consistent and accurate seizure predictability across seizures in the same patient and across patients. The null hypothesis we considered is that the obtained average value of T_p from the optimal sites across all seizures is statistically smaller or equal to the average T_p from randomly selected ones. T_p values were obtained for a total of 10 randomly selected tuples of five sites per seizure per measure. For every measure, if the T_{p-opt} values (T_p values obtained from optimal electrode sites) were greater than 2.58 standard deviations from the mean of the $T_{p-random}$ (T_p values obtained from randomly selected electrode sites) values, the null hypothesis was rejected at $\alpha = 0.01$ (262 degrees of freedom for patient 1, that is 24 seizures × 10 random tuples of sites − 1 + 24 seizures × 1 optimal tuple of sites − 1; and similarly 207 degrees of freedom for patient 2). The T_{p-opt} values were significantly larger than the $T_{p-random}$ values for all the three T-index based measures (see Table 3). This result was consistent across both patients and further supports the hypothesis that the spatiotemporal dynamics of synchronization of critical (optimal) brain sites per synchronization measure should be followed in time to observe significant preictal changes predictive of an upcoming seizure. These results also imply the importance of the use of spatiotemporal synchronization via multi-electrode analysis (in both space and time) for seizure prediction.

Table 3 Statistical significance values for tests of brain resetting using LR and RR

Patient	LR			RR		
	$\hat{p}^*_{LR}$	$\hat{p}_{LR}$ (μ, σ)	P value	$\hat{p}^*_{RR}$	$\hat{p}_{RR}$ (μ, σ)	p value
1	0.75 (18/24)	$\mu = 0.46$ $\sigma = 0.17$	$P < 0.05$	0.875 (21/24)	$\mu = 0.41$ $\sigma = 0.12$	$p < 0.0005$
2	0.63 (12/19)	$\mu = 0.39$ $\sigma = 0.14$	$P < 0.05$	0.78 (15/19)	$\mu = 0.42$ $\sigma = 0.13$	$p < 0.005$
Overall	0.69 (30/43)	$\mu = 0.42$ $\sigma = 0.15$	$p < 0.05$	0.83 (36/43)	$\mu = 0.42$ $\sigma = 0.13$	$p < 0.003$

3 Brain Resetting

We have found that seizures are not abrupt transitions in and out of an abnormal ictal (seizure) state, instead they follow a dynamical transition that evolves over minutes to hours [22, 23]. During this preictal dynamical transition, multiple regions of the cerebral cortex progressively approach a similar dynamical state. The dynamics of the preictal transition are highly complex. Even in the same patient, the participating cortical regions and the duration of the transition vary from seizure to seizure. Understanding these processes requires analytic methods capable of identifying anatomical regions where critical state changes occur. During the seizure (ictal state), widespread cortical areas make an abrupt transition to a more ordered state. After the seizure, brain dynamics revert to a more disordered state in which previously entrained (synchronized) cortical areas become disentrained (desynchronized – postictal state). The epileptic brain repeats this series of state transitions intermittently, at seemingly irregular but, in fact, time-dependent intervals [17, 20]. This implies that the transition into seizures is not a random process. Thus, the question as to why seizures occur should be addressed. Based on the analysis of a small sample of seizure recordings, we postulated that seizures may serve as intrinsic mechanisms to desynchronize brain areas that were dynamically synchronized in the immediate preictal periods [40, 41, 74]. We have defined this phenomenon as "resetting" of the epileptic brain via recurrent seizures [40]. Figure 4 shows the dynamical progressive resetting of the brain via two seizures. From this figure, we observe that when the first seizure failed to reset its preceding preictal synchronization, a second seizure occurs that did desynchronize the brain. Elucidation of the mechanisms of this resetting process is a challenging task. In this section, we describe two characteristics of this resetting process: (a) *Level of Resetting (LR)*, quantified by the difference of the average T-index value at critical sites between 10 min before (immediate preictal state) and 10 min after (immediate postictal state) a seizure's onset; (b) *Resetting Ratio of Time constants (RRT)*, quantified by the ratio of the time it takes the critical sites to disentrain postictally (desynchronization period) over the time they were entrained preictally (synchronization period). This ratio is equivalent to the estimation of the ratio of the slope of the T-index curve in the postictal over the one in the preictal period, because the synchronization/desynchronization

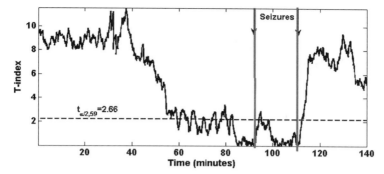

Fig. 4 Resetting of EEG by multiple seizures. T-index profile of a pair of electrodes that are synchronized in STL_{max} prior to the first seizure. They did not desynchronize postictally. Within 18 min after the first seizure, a second seizure occurs and desynchronizes this electrode pair. *Red* vertical lines denote the seizures' onset. Typically, a seizure lasted for 2–3 min

threshold for the T-index values is the same ($T_{th} = 2.662$) – the end points of the synchronization and desynchronization periods correspond to the same T-index value T_{th}. Using these two metrics of resetting, we demonstrate that brain resetting of dynamically synchronized cortical sites, as measured by *LR* or *RRT*, occurs much more frequently at seizure points than at randomly selected points in the available interictal EEG recordings per patient. The implications of these findings for the understanding of epileptogenesis, seizure prediction and control are discussed. Also, in the following subsections, for the analysis of the phenomenon of resetting, we use the T-index on the STL_{max} profiles, that is the profiles that showed the most promising results for seizure predictability.

3.1 Resetting Characteristics

3.1.1 Level of Resetting (LR)

One way to quantify the observed differences in the level of synchronization of the critical sites around seizure points is to estimate a statistically significant change between their synchronization and desynchronization levels, as these are estimated by their corresponding average T-index profile. Thus, the *maximum synchronization level* ($L_{syn}(t^*)$) is defined as the minimum average T-index value of the sites selected within the 10 min preictal window $w_1(t^*)$. In analogy, the *maximum desynchronization level* ($L_{desyn}(t^*)$) is defined as the average T-index value of the preictally selected sites within a 10 min postictal window after the end of a seizure, that is $w_1(t^* + \text{seizure duration} + 10\,\text{min})$. (Note that $w_1(t)$ is always moved forward in time.) The hypothesis was that higher levels of preictal synchronization (marked by low values of T-index) are reduced significantly during the corresponding postictal period (marked by high values of T-index), and that this phenomenon occurs more often around seizure points t^* than around any other point t of the recording.

Criterion for resetting based on LR: We define a resetting occurring at time t if the difference $L_{\text{desyn}}(t) - L_{\text{syn}}(t) > 2.662$. If p_{LR} is the probability of resetting at a time point t, and we denote by x_l the random variable of the number of resettings observed at l time points, x_l follows the binomial (l, p_{LR}) distribution, where p_{LR} can be estimated by the maximum likelihood estimate $\hat{p}_{LR}$ as

$$\hat{p}_{LR} = (\text{number of time points t where} L_{\text{desyn}}(t) - L_{\text{syn}}(t) > 2.662)/l = x_l/l$$

The $\hat{p}_{LR}$s are estimated from l randomly (following a uniform random distribution) chosen time points t (forming l-tuple of time points) within the available time interval of recordings from each patient, excluding ictal periods (on average 2.5 min ictal duration per seizure for the two analyzed patients). A Monte-Carlo estimation of the probability distribution of $\hat{p}_{LR}$ is produced by generating the histogram of $\hat{p}_{LR}$ from 25,000 l-tuples (l differs from patient to patient, being equal to the number of recorded seizures per patient). The $\hat{p}_{LR}$ at seizure points t^*, denoted by $\hat{p}^*_{LR}$, is then compared to the distribution of $\hat{p}_{LR}$s at time points t to statistically decide if $\hat{p}^*_{LR} > \hat{p}_{LR}$.

3.1.2 Resetting Ratio of Time Constants (RRT)

One way to quantify the observed differences in the rates of synchronization of the critical sites around seizure points t^* is to estimate the synchronization and desynchronization periods from the corresponding average T-index profile, using the same statistical threshold $T_{th} = 2.662$ for both periods. Thus, the *synchronization period* (T_{syn}) is defined as the period before a time point t, during which the average T-index values of the sites selected from the 10 min (60 points) window $w_1(t)$ remain continuously below 2.662. For seizure points, $t = t^*$ and $T_{\text{syn}} = T_p$ (prediction time) In analogy, the *desynchronization period* (T_{desyn}) was defined as the period after time $t + 2.5$ min (allowing for a virtual seizure duration of 2.5 min) during which the average T-index values of the selected sites remain continuously below 2.662. The hypothesis was that preictal entrainment lasts significantly longer than the corresponding postictal disentrainment and, in accordance with the previously observed phenomenon of hysteresis and the theory of nonlinear dynamics at points of critical transitions, this phenomenon occurs more often at seizure points than at any other time points. This phenomenon is consistent with what is called "critical slowing down" in the literature of the dynamics of phase transitions [9, 25].

Criterion for resetting based on RRT: We define a resetting occurring at time t if $T_{\text{syn}}(t) < T_{\text{desyn}}(t)$. If p_{RRT} is the probability of such a resetting at a time point t, and we denote by x_l the random variable of the number of such resettings observed at l time points, x_l follows the binomial (l, p_{RRT}) distribution, where p_{RRT} can be estimated by the maximum likelihood estimate $\hat{p}_{RRT}$ as

$$\hat{p}_{RRT} = (\text{number of time points t where} T_{\text{syn}}(t) < T_{\text{desyn}}(t))/l = x_l/l$$

The $\hat{p}_{RRT}$s are estimated from l randomly (following a uniform random distribution) chosen time points t (forming l-tuple of time points) within the available time interval of recordings from each patient, excluding ictal periods as before. A Monte-Carlo estimation of the probability distribution of $\hat{p}_{RRT}$ is produced by generating the histogram of $\hat{p}_{RRT}$ from 25,000 l-tuples. The $\hat{p}_{RRT}$ at seizure points t^*, denoted by $\hat{p}^*_{RRT}$, is then compared to the distribution of $\hat{p}_{RRT}$s at time points t to statistically decide if $\hat{p}^*_{RRT} > \hat{p}_{RRT}$.

Both null resetting hypotheses above were tested and rejected (see next section) – assuming that the involved distributions were binomial. Therefore, it appears that resetting of the epileptic brain at seizures manifests itself as a reversal of the pathology of excessive entrainment (synchronization), and that this reversal occurs mostly at seizures than at other time points.

3.2 Resetting Results

We tested the following two null hypotheses: (1) H_0: $\hat{p}^*_{LR} = \hat{p}_{LR}$ versus the alternative hypothesis H_a: $\hat{p}^*_{LR} > \hat{p}_{LR}$, and 2) H_0: $\hat{p}^*_{RRT} = \hat{p}_{RRT}$ versus the alternative hypothesis H_a: $\hat{p}^*_{RRT} > \hat{p}_{RRT}$. This is performed by comparing the observed $\hat{p}^*$ with the distribution of the $\hat{p}$ values at time points in non-seizure periods, for each of the two criteria of resetting. Then, at the 95% confidence level, we can reject the two H_0 if the p-value of the tests is less than 0.05. For each of the 2 patients, the probability of resetting at the l seizure points, using either of the two criteria (overall 69% for LR, 83% for RRT), was significantly greater than the corresponding one at l points randomly selected in interictal periods, where resetting is observed with a 38% and 40% probability for the criteria of LR and RRT respectively (see mean value μ in second and fifth columns in Table 3). The most significant test for resetting at seizures appears to be the RRT criterion with a p-value well less than 0.01. Also, of note here is that, according to the binomial distribution, an expected value of 50% for the probability of resetting is anticipated in the case of a random process. Therefore, the finding of a reduced overall probability of approximately 40% of resetting using either of the resetting criteria during the interictal periods implies that the brain resets less than randomly during the interictal periods. We believe that such an observation further supports our previous findings that seizures are characterized by a progressive spatiotemporal entrainment of cerebral sites that may last for minutes to hours to days, which implies that interictal periods may simply be intermittent states (in the way this term is used in nonlinear dynamics) than normal ones.

4 Modeling of Epileptic Seizures

The preictal dynamical synchronization (entrainment) and postictal synchronization (disentrainment) of $STL_{\max}$ of critical sites has already constituted the basis for (a) the design and development of systems for long-term (tens of minutes), on-line, prospective prediction of epileptic seizures [63–65, 75–77], and (b) the detection of seizure susceptibility periods on the order of hours to days prior to a seizure occurrence [24]. We will show that the above observations are in agreement with corresponding ones in systems of coupled neural population models with internal (i.e., part of the overall system) pathological feedback. By increasing the interpopulation coupling and detuning their internal feedback mechanism, these systems move from spatiotemporal chaotic states into spatially synchronized, quasiperiodic, "seizure-like" states. These transitions are characterized by spatial convergence of the maximum Lyapunov exponents of the individual populations. These results address ictogenesis and appear to support the hypotheses that: (a) seizures may result from the inability of internal feedback mechanisms to provide timely compensation/regulation of coupling between brain sites, (b) seizure control can be achieved by feedback decoupling of the "pathological" sites via external stimuli.

4.1 Neural Population Model

To show that the above hypotheses are plausible, we herein use a coupled cortical population model (CM), which is based on the thalamocortical neural population model by Suffczynski et al. [78]. From this thalamocortical model, we retained the cortical module which consists of 2 cell subpopulations: pyramidal cells (PY) and cortical interneurons (IN). The interaction between these subpopulations is via AMPA-mediated (AMPA: α-amino-3-hydroxy-5-methyl-4-isoxazolepropionic acid) excitatory synapses, and $GABA_A$ and $GABA_B$-mediated (GABA: γ-aminobutyric acid) inhibitory synapses. The schematic diagram of the CM population is shown in Fig. 5a.

Each neural subpopulation is described via two variables – membrane potential and firing rate. The synaptic transmission (i.e., firing rate to membrane potential conversion) is described by the following equation of the transmembrane potential $V(t)$ for neurons in each of the above two subpopulations.

$$C_m \dot{V}(t) = -\Sigma I_{syn} - g_{leak}(V(t) - V_{leak}) \tag{7}$$

where the synaptic currents due to AMPA, $GABA_A$ and $GABA_B$ receptors are $I_{syn}(t) = g_{syn}(t)(V(t) - V_{syn})$, C_m is the membrane capacitance, g_{leak}, $g_{syn}(t)$ are the leak and synaptic current conductances, and V_{leak}, V_{syn} are the corresponding (reversal) potentials. The synaptic conductances are obtained by *convolving* the incoming action potential firing rate $F(t)$ with the synaptic impulse response $h_{syn}(t) = A_{syn}(e^{a_1 t} - e^{a_2 t})$ where $a_2 > a_1$ (fast and slow synaptic kinetics). Different values for V_{syn}, A_{syn}, a_1 and a_2 are used to model synaptic transmission mediated by AMPA, $GABA_A$ and $GABA_B$. $GABA_B$ receptors that have nonlinear activation properties (see [78] for details). A sigmoidal static nonlinearity is used to transform back the membrane potentials into firing rates:

Brain Dynamics and Modeling in Epilepsy: Prediction and Control Studies

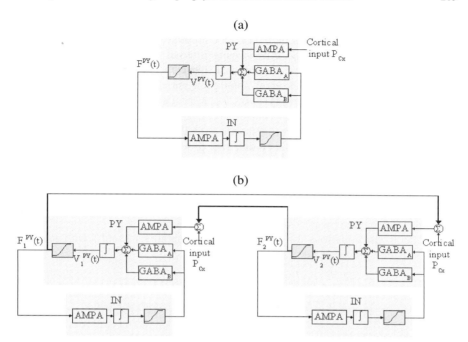

Fig. 5 Cortical neural population model. (**a**) Block diagram of a single CM population, with PY and IN denoting the pyramidal and interneuron subpopulations, and $P_{Cx}(t)$, $V^{PY}(t)$ the input and output signals from and to the CM respectively. (**b**) Block diagram of two coupled CMs (CCM). The pyramidal cell average membrane potentials ($V_1^{PY}(t)$, $V_2^{PY}(t)$) are the observed outputs from the CCM. The AMPA-mediated excitatory interactions ($F_1^{PY}(t)$, $F_2^{PY}(t)$) between CM populations (bold lines on top) are via pyramidal cells only, and represent inputs to the CMs, in addition to the $P_{Cx}(t)$ ones

$$F(t; V(t)) = \frac{G_s}{1 + e^{(V(t) - \Theta_s)/\zeta_s}} \qquad (8)$$

where G_s is the maximal firing rate, and Θ_s, ζ_s are the sigmoid threshold and slope parameters. The intra-CM interactions, i.e. the ones between the PY and IN subpopulations, are modeled using constant coupling gains. Cortical input $P_{Cx}(t)$ is modeled by non-zero mean white Gaussian noise. For appropriate value of $\langle P_{Cx} \rangle$ (mentioned below), where $\langle . \rangle$ denotes signal mean, the CM exhibits "normal" activity. The mean membrane potential of the pyramidal cells $V^{PY}(t)$ is the observed model output, which simulates experimental recordings of local field potentials. We refer the reader to [78] for further implementation details.

We model the inter-population coupling exclusively with excitatory connections between pyramidal cells, which is consistent with neurophysiology [79]. A single synaptic gain factor represents the coupling as follows. Let $F_1^{PY}(t)$, $F_2^{PY}(t)$ (see Fig. 5b) correspond to the firing rates of pyramidal cells in CM populations 1 and 2 respectively. The coupling from population 2 to population 1 is $K_{2 \to 1} F_2^{PY}(t)$.

Similarly, the coupling from population 1 to population 2 is $K_{1\to 2} F_1^{PY}(t)$. (For simplicity, henceforth we use equal bidirectional coupling, i.e., $K_{2\to 1}(t) = K_{1\to 2}(t) = K(t)$, where $K(t)$ represents the time-varying internal coupling strength that we externally modify at will in this simulation experiment.) The coupling signal affects the pyramidal cells in the target population via excitatory AMPA-mediated synapses. For the sake of generality, two CMs were made non-identical by setting $\langle P_{Cx} \rangle$ at different values (8 pps – pulses per second, and 10 pps in the respective CMs in our simulations here), and with "small" perturbations to the AMPA, GABAA, GABAB amplitude, decay and rise times, and to the intra-population coupling gains (Note that similar results to the ones we will report here were also obtained for several different parameterizations). All other parameters are similar to [78]. Then, an uncoupled neural population is generally described by $\dot{V}_i(t) = f_i(V_i(t), F_i(t))$, where $V_i(t)$ and $F_i(t)$ are the i-population's membrane potential and firing rate respectively.

In general, the equations for coupled neural populations i and j with internal feedback regulation would be

$$\dot{V}_i(t) = f_i\left(V_i(t), F_i(t)\right) + (K(t) - \varepsilon(t)) T\left(F_j(t)\right)$$
$$\dot{V}_j(t) = f_j\left(V_j(t), F_j(t)\right) + (K(t) - \varepsilon(t)) T\left(F_i(t)\right) \quad (9)$$

where $T(.)$ denotes the synaptic conversion of the afferent firing rate to membrane potential, $K(t)$ denotes the changes in excitatory coupling, while $\varepsilon(t)$ is the internal feedback term such that the effective coupling $K(t) - \varepsilon(t)$ maintains normal synchronization levels between the populations i and j. This internal feedback can be modeled as a proportional-integral (PI) feedback [81]:

$$\varepsilon(t) = PI_I(\rho_{i,j}(t) - c_*) \quad (10)$$

where PI_I denotes the proportional-integral action, $\rho_{i,j}(t)$ is the level of synchronization between i and j, and c_* is a threshold for "normal" synchronous activity. The subscript I in PI_I denotes that the feedback is an internal mechanism of the "brain". (The internal feedback is not explicitly shown in Fig. 5b). For simulation expedience, the synchronization level is quantified by the estimate of the cross-correlation coefficient $\rho_{1,2}(t)$ between the two neural population outputs, and it is computed in an exponentially weighted manner [44, 45]. Despite the highly nonlinear nature of the neural population models involved, it is observed that a simple PI compensator is sufficient to achieve their decorrelation, as long as the closed-loop bandwidth is not too large. For tuning the PI we followed [80]. The PI_I is restricted to produce signals ≥ 0 and it employs limited integration as an anti-windup mechanism (e.g., see [81]). This guarantees that when the correlation between the two signals is below the threshold c_* ("normal" synchrony), no PI feedback is generated. The PI_I generates an output that attempts to cancel the effect of the excessive coupling in the neural network and maintains the correlation between the respective two signals below a given threshold c_*. Thus, from a feedback point of view, the PI counteracts the destabilizing effect of excessive excitatory coupling. During "normal" behavior, $\varepsilon(t)$ follows changes in $K(t)$ and regulates it when $\rho(t) > c_*$. For simulations here, we set $c_* = 0.1$. In order to model "pathological" feedback in the neural populations, the PI_I is de-tuned. For example, to simulate pathological internal regulation, we use a "pathological" (low-gain) $PI_I = 0.004 + 0.0009/s$ for the CCM. Results from de-tuned (pathological) internal feedback action are shown in Fig. 6. The detuned PI_I is out of phase with the coupling change $K(t)$ and the effective coupling $K(t) - \varepsilon(t)$ keeps increasing. When $K(t) - \varepsilon(t)$ exceeds a critical value $K_c(t)$ (which depends on the transfer functions used to model the neural populations), it "destabilizes" the neural network, causing its output to grow. This occurrence could be seen from a bifurcation analysis of the coupled systems, where the overall nonlinear system exhibits stable self-sustained oscillations (high-amplitude "bursts").

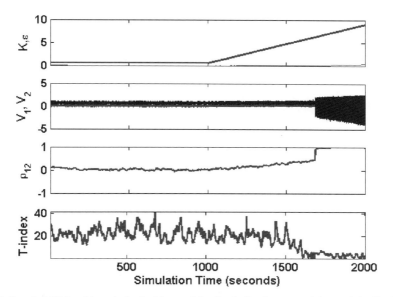

Fig. 6 Coupled CMs with pathological internal feedback. During normal internal feedback, the correlation between the neural populations remains low throughout (i.e., $\rho_{1,2}(t) \leq 0.1$) despite the increase of coupling $K(t)$, since normal internal feedback can track these changes and maintain "normal" correlation by constraining the effective coupling $K(t) - \varepsilon(t)$. With reduced (pathological) internal feedback gain, the internal regulator can no longer follow coupling changes closely. "Seizure-like" bursts appear soon after the net coupling $K(t) - \varepsilon(t)$ exceeds a threshold $K_c(t)$ ($K_c(t) \approx 7$). Notice the significant increase in signal correlation and the T-index of the outputs of the two populations preceding the "seizures", that bears similarities to the entrainment (synchronization) observed in actual epileptic EEG prior to seizures. Panel legends (*top to bottom*): **Panel I.** Coupling $K(t)$ and regulating internal feedback gain $\varepsilon(t)$. **Panel II.** Correlation between population outputs. **Panel III.** Population output signals. **Panel IV.** The T-index between the $STL_{\max}$ of the output signals obtained from the two populations

5 Control of Epileptic Seizures

Conceptually, the excess excitatory synaptic coupling can be removed by subtracting an estimate of the actual coupling signal, or be compensated by an inhibitory control signal. We have called such a controlling scheme "feedback decoupling" [44]. The implementation aspects of the "subtractive" feedback decoupling control of CCMs, having stimulation currents as control inputs, is as follows. With an external stimulation current $u_E(t)$ (e.g. via neurostimulation electrodes), the pyramidal cell subpopulation average membrane potential is described by $\dot{V}_{PY}(t) = -\Sigma I_{int}(t) + u_E(t)$ (see Fig. 7 and Eq. (7)), where $-\Sigma I_{int}(t)$ is the sum of intra-population and inter-population synaptic and leak currents. The effect of excessive excitatory synaptic coupling current from population 2 to population 1 can be removed with:

$$u_{E,1}(t) = -K_E(t)\left(\hat{h}_{AMPA}(t)^* \hat{F}_2^{PY}(t)\right)\left(V_1^{PY}(t) - V_{AMPA}\right) \quad (11)$$

and similarly for population 2. $\hat{h}_{AMPA}(t)$ is an estimate of the AMPA synaptic kinetics, V_{AMPA} is the synaptic reversal potential, $\hat{F}^{PY}(t)$ is an estimate of the coupling signal (pyramidal cell firing rate) convolved (*) with $\hat{h}_{AMPA}(t)$, and $K_E(t)$ is the externally manipulated control gain. $\hat{h}_{AMPA}(t)$ and V_{AMPA} can be obtained empirically. Here, $\hat{h}_{AMPA}(t)$ is approximated with a low-pass filter (bandwidth 20 Hz). The accuracy of the synaptic kinetics (for example, rise and settling time) is not critical. $\hat{F}_i^{PY}(t)$ is estimated from the observed signals $V_i^{PY}(t)$ via a rectifier, i.e., $\hat{F}_i^{PY}(t) = V_i^{PY}(t)$ if $V_i^{PY}(t) \geq 0$; $\hat{F}_i^{PY}(t) = 0$ otherwise. This approximation of the nonlinear conversion of membrane potential to action potential firing rate turns out to be sufficient for feedback decoupling control. The gains $K_E(t)$ are manipulated by an external PI controller (PI_E), i.e., $K_E(t) = PI_E(\rho_{i,j}(t) - c_*)$, where $PI_E = 1.7 + 0.44/s$.

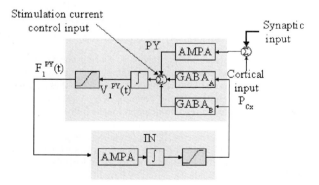

Fig. 7 Feedback decoupling control. This figure shows one of the two coupled CMs with the externally applied stimulation current control input $(u_E(t))$. Different control inputs control the pyramidal cell population in each CM

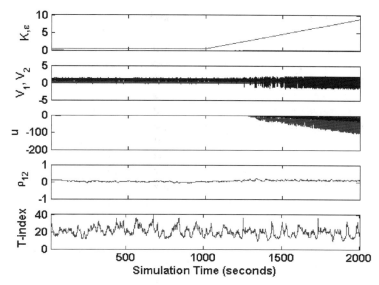

Fig. 8 Successful closed-loop control of an epileptic CCM via feedback decoupling by current stimulation of the pyramidal subpopulations of the two CM populations. The excess internal excitatory coupling developed due to low (pathological) internal feedback at high values of internal coupling is removed by subtracting its estimate through the external controller. *Top to bottom panels*: **Panel I**. Internal coupling $K(t)$ (large) and internal feedback coupling $\varepsilon(t)$ (very small, almost zero). **Panel II**. The output signals from the two populations. **Panel III**. The externally supplied currents $u_{E,1}(t)$ and $u_{E,2}(t)$ to the two populations in μA. **Panel IV**. Correlation between system outputs remains close to zero. **Panel V**. T-index between the STL_{max} of the output signals from the two populations remains very high over time, denoting very low dynamical synchronization of the two CM populations of the CCM despite the increase in the interpopulation coupling $K(t)$

Similar to the internal feedback, PI_E provides feedback stimulation when $\rho_{i,j}(t) > c_*$ to compensate for the lack of appropriate internal feedback. Successful "seizure" control is shown in Fig. 8, wherein "normal" correlation is maintained throughout by the external feedback decoupling controller. An advantage of the use of such a conceptual feedback controller to decouple two populations is that the requirements on model accuracy are less stringent and effective than those for open-loop control [44]. Control of "seizures" was possible with the same external control stimulation setup for various intra-population system parameters. Excess excitatory synaptic coupling could also be compensated with inhibitory stimulation current, which resembles GABAA-mediated synaptic coupling [45–47].

6 Conclusions

It was not many years ago that epileptic seizures were considered random events that could not be anticipated. The results presented above, as well as other published over the last years, show that this is not the case. It appears that design of seizure prediction algorithms with performance similar to the seizure predictability results reported herein might be feasible and worth trying. Sensitivity, specificity and prediction time of such algorithms have to be carefully investigated in real-time, prospectively and on-line in long, continuous EEG recordings across many patients, as well as many seizures in the same patient. The developed concept of seizure resetting looks counter-intuitive, especially under the currently prevailing theory of "seizures beget seizures". However, we showed that this is indeed the case when one examines the phenomenon from a dynamical point of view: the brain dynamics reset more at seizures than at any other point in time. The third concept we introduced in this study has to do with the seizure control, and it inherently also suggests a model for generation of seizures (ictogenesis). In particular, while the concept of seizure predictability was based on the hypothesis of preictal spatial synchronization of the brain dynamics, and the seizure resetting on postictal spatial desynchronization, the seizure control was based on the hypothesis of the existence of pathological internal feedbacks in the epileptic brain that cannot compensate for intermittent spatial hypersynchronization of dynamics. All three concepts are interrelated and may provide the key ingredients for the development of robust brain pacemakers for the treatment of epilepsy in the near future, within the framework of neuromodulation.

Acknowledgments A. Prasad acknowledges the support by the DST, India.

References

1. J. Engel Jr., P. C. Van Ness, T. B. Rasmussen and L. M. Ojemann, "Outcome with respect to epileptic seizures," in *Surgical Treatment of the Epilepsies*, J. Engel Jr., Ed., New York: Raven Press, pp. 609–622, 1993.
2. J. Engel Jr., *Seizures and Epilepsy*. Philadelphia, PA: F. A. Davis Co., 1989.
3. J. Engel Jr. and T. A. Pedley, *Epilepsy: a Comprehensive Textbook*. Philadelphia, PA: Lippincott-Raven, 1997.
4. J. G. Milton, J. Gotman, G. M. Remillard and F. Andermann, "Timing of seizure recurrence in adult epileptic patients: a statistical analysis," *Epilepsia*, vol. 28, pp. 471–478, 1987.
5. W. G. Lennox, *Science and Seizures*. New York: Harper, 1946.
6. W. Penfield, "The evidence for a cerebral vascular mechanism in epilepsy," *Ann. Int. Med.*, vol. 7, pp. 303–310, 1933.
7. F. M. Forster, *Reflex Epilepsy, Behavioral Therapy and Conditioned Reflexes*. Springfield: Thomas, 1977.
8. C. D. Binnie, A. J. Wilkins, F. C. Valdivia, F. J. J. Jimenez, J. Tejeiro, L. A. Peralta, A. Vaquero and E. G. Albea, "Ecstatic seizures by television," *J. Neurol. Neurosur. Psychiat.*, vol. 63, p. 273, 1997.
9. H. Degn, A. Holden and L. F. Olsen, *Chaos in Biological Systems*. New York: Plenum, 1987.
10. W. J. Freeman, "Simulation of chaotic EEG patterns with a dynamic model of the olfactory system," *Biol. Cybern.*, vol. 56, pp. 139–150, 1987.
11. A. Babloyantz and A. Destexhe, "Low dimensional chaos in an instance of epilepsy," *Proc. Natl. Acad. Sci. USA*, vol. 83, pp. 3513–3517, 1986.
12. G. W. Frank, T. Lookman, M. A. H. Nerenberg, C. Essex et al., "Chaotic time series analysis of epileptic seizures," *Physica D*, vol. 46, pp. 427–438, 1990.

13. J. C. Principe, A. Rathie and J. M. Kuo, "Prediction of chaotic time series with neural networks and the issue of dynamical modeling," *Int. J. Bifurcat. Chaos*, vol. 2, pp. 989–996, 1992.
14. L. D. Iasemidis, H. P. Zaveri, J. C. Sackellares, W. J. Williams and T. W. Hood, "Nonlinear dynamics of electrocorticographic data," *J. Clin. Neurophysiol.*, vol. 5, p. 339, 1988.
15. L. D. Iasemidis, H. P. Zaveri, J. C. Sackellares and W. J. Williams, "Linear and nonlinear modeling of ECoG in temporal lobe epilepsy", in *Proceedings of the 25th Annual Rocky Mountain Bioengineering Symposium*, vol. 24, pp. 187–193, 1988.
16. L. D. Iasemidis, J. C. Sackellares, H. P. Zaveri and W. J. Williams, "Phase space topography of the electrocorticogram and the Lyapunov exponent in partial seizures," *Brain Topogr.*, vol. 2, pp. 187–201, 1990.
17. L. D. Olson, L. D. Iasemidis and J. C. Sackellares, "Evidence that interseizure intervals exhibit low dimensional dynamics," *Epilepsia*, vol. 30, p. 644, 1989.
18. L. D. Iasemidis, J. C. Sackellares and W. J. Williams, "Localizing preictal temporal lobe spike foci using phase space analysis," *Electroencephalogr. Clin. Neurophysiol.*, vol. 75, pp. S63–S64, 1990.
19. L. D. Iasemidis, J. C. Sackellares and R. S. Savit, "Quantification of hidden time dependencies in the EEG within the framework of nonlinear dynamics", in *Nonlinear Dynamical Analysis of the EEG*, B. H. Jansen and M. E. Brandt, Eds., Singapore: World Scientific, pp. 30–47, 1993.
20. L. D. Iasemidis, L. D. Olson, J. C. Sackellares and R. Savit, "Time dependencies in the occurrences of epileptic seizures: a nonlinear approach," *Epilepsy Res.*, vol. 17, pp. 81–94, 1994.
21. J. C. Sackellares, L. D. Iasemidis, A. Barreto, R. L. Gilmore, R. S. Savit, B. M. Uthman and S. N. Roper, "Computer-assisted seizure detection based on quantitative dynamical measures," *Electroenceph. Clin. Neurophysiol.*, vol. 95(2), p. 18P, 1995.
22. L. D. Iasemidis and J. C. Sackellares, "The temporal evolution of the largest Lyapunov exponent on the human epileptic cortex," in *Measuring Chaos in the Human Brain*, D. W. Duck and W. S. Pritchard, Eds., Singapore: World Scientific, pp. 49–82, 1996.
23. L. D. Iasemidis, J. C. Principe, J. M. Czaplewski, R. L. Gilman, S. N. Roper and J. C. Sackellares, "Spatiotemporal transition to epileptic seizures: A nonlinear dynamical analysis of scalp and intracranial EEG recordings," in *Spatiotemporal Models in Biological and Artificial Systems*, F. L. Silva, J. C. Principe and L. B. Almeida, Eds., Amsterdam: IOS Press, pp. 81–88, 1997.
24. J. C. Sackellares, L. D. Iasemidis, R. L. Gilmore and S. N. Roper, "Epilepsy – when chaos fails," in *Chaos in the Brain?* K. Lehnertz, J. Arnhold, P. Grassberger and C. E. Elger, Eds., Singapore: World Scientific, pp. 112–133, 2000.
25. L. D. Iasemidis, "Epileptic seizure prediction and control", *IEEE Transactions on Biomedical Engineering*, vol. 50(5), pp. 549–558, 2003.
26. K. Lehnertz and C. E. Elger, "Spatio-temporal dynamics of the primary epileptogenic area in temporal lobe epilepsy characterized by neuronal complexity loss," *Electroencephalogr. Clin. Neurophysiol.*, vol. 95, pp. 108–117, 1995.
27. K. Lehnertz and C. E. Elger, "Neuronal complexity loss of the contralateral hippocampus in temporal lobe epilepsy: a possible indicator of secondary epileptogenesis," *Epilepsia*, vol. 36(S4), p. 21, 1995.
28. D. A. Scott and S. J. Schiff, "Predictability of EEG interictal spikes," *Biophys. J.*, vol. 69, pp. 1748–1757, 1995.
29. F. H. Lopes da Silva, J. P. Pijn and W. J. Wadman, "Dynamics of local neuronal networks: control parameters and state bifurcations in epileptogenesis," *Prog. Brain Res.*, vol. 102, pp. 359–370, 1994.
30. L. D. Iasemidis, J. C. Sackellares, Q. Luo, S. N. Roper and R. L. Gilmore, "Directional information flow during the preictal transition," *Epilepsia*, vol. 40(S7), pp. 165–166, 1999.
31. L. D. Iasemidis, J. C. Principe and J.C. Sackellares, "Measurement and quantification of spatiotemporal dynamics of human epileptic seizures," in *Nonlinear Biomedical Signal Processing*, M. Akay, Ed., IEEE Press, vol. II, pp. 294–318, 2000.

32. J. C. Sackellares, L. D. Iasemidis, R. L. Gilmore and S. N. Roper, "Epilepsy – when chaos fails" in *Chaos in the Brain?*, K. Lehnertz, J. Arnhold, P. Grassberger and C. E. Elger, Eds., Singapore: World Scientific, pp. 112–133, 2000.
33. L. D. Iasemidis, D. S. Shiau, P. Pardalos and J. C. Sackellares, "Transition to epileptic seizures – an optimization approach into its dynamics", in *Discrete Problems with Medical Applications*, D. Z. Du, P. M. Pardalos and J. Wang, Eds., DIMACS series, American Mathematical Society Publishing Co., vol. 55, pp. 55–74, 2000.
34. L. D. Iasemidis, P. Pardalos, J. C. Sackellares and D. S. Shiau, "Quadratic binary programming and dynamical system approach to determine the predictability of epileptic seizures," *J. Comb. Optim.*, vol. 5, pp. 9–26, 2001.
35. L. D. Iasemidis, P. M. Pardalos, J. C. Sackellares and V. Yatsenko, "Global optimization approaches to reconstruction of dynamical systems related to epileptic seizures," in *Mathematical Methods in Scattering Theory and Biomedical Technology*, C. V. Massalas et al., Eds., Singapore: World Scientific, pp. 308–318, 2002.
36. L. D. Iasemidis, D. S. Shiau, P. M. Pardalos and J. C. Sackellares, "Phase Entrainment and Predictability of Epileptic Seizures," in *Biocomputing*, P. M. Pardalos and J. Principe, Eds., Kluwer Academic Publishers, pp. 59–84, 2002.
37. L. M. Hively, N. E. Clapp, C. S. Daw and W. F. Lawkins, "Nonlinear analysis of EEG for epileptic events," ORNL/TM-12961, Oak Ridge National Laboratory, Oak Ridge, TN, 1995.
38. B. Litt, R. Esteller, J. Echauz, M. D. Alessandro, R. Shor, T. Henry, P. Pennell, C. Epstein, R. Bakay, M. Dichter and G. Vachtsevanos, "Epileptic seizures may begin hours in advance of clinical onset: A report of five patients," *Neuron*, vol. 30, pp. 51–64, 2001.
39. L. D. Iasemidis, D. S. Shiau, W. Chaovalitwongse, P. M. Pardalos, P. R. Carney and J. C. Sackellares "Adaptive seizure prediction system," *Epilepsia*, vol. 43, pp. 264–265, 2002.
40. L. D. Iasemidis, D. S. Shiau, J. C. Sackellares, P. M. Pardalos and A. Prasad, "Dynamical resetting of the human brain at epileptic seizures: application of nonlinear dynamics and global optimization techniques," *IEEE Trans. Biomed. Eng.*, 51(3), pp. 493–506, 2004.
41. L. D. Iasemidis, A. Prasad, J. C. Sackellares, P. M. Pardalos and D.-S. Shiau, "On the prediction of seizures, hysteresis and resetting of the epileptic brain: insights from models of coupled chaotic oscillators," in *Order and Chaos*, T. Bountis Ed. vol. 8, Publishing House of K. Sfakianakis, Thessaloniki: Greece, in press (Proceedings of the 14th Summer School on *Nonlinear Dynamics: Chaos and Complexity*, Patras, Greece, 2001).
42. K. Tsakalis, "Performance limitations of adaptive parameter estimation and system identification algorithms in the absence of excitation," *Automatica*, vol. 32, pp. 549–560, 1996.
43. K. S. Tsakalis, "Bursting scenaria in adaptive algorithms: Performance limitations and some remedies," *Kybernetika*, vol. 33, pp. 17–40, 1997.
44. K. S. Tsakalis, L. D. Iasemidis, "Control aspects of a theoretical model for epileptic seizures," *Int. J. Bif. Chaos*, vol. 16, pp. 2013–2027, 2006.
45. K. Tsakalis, N. Chakravarthy, S. Sabesan, L. D. Iasemidis, and P. M. Pardalos, "A feedback control systems view of epileptic seizures," *Cybernetics Sys. Anal.*, vol. 42, pp. 483–495, 2006.
46. N. Chakravarthy, S. Sabesan, L. D. Iasemidis, and K. Tsakalis, "Modeling and controlling synchronization in a neuron-level population model," *Int. J. Neural Sys.*, vol. 17(2), pp. 123–38, 2007.
47. N. Chakravarthy, S. Sabesan, L. D. Iasemidis, and K. Tsakalis, "Controlling epileptic seizures in a neural mass model," *J. Comb. Optim.*, 2008 (in press).
48. M. Harrison et al., "Accumulated energy revisited" *Clinical Neurophysiol.*, vol. 116, pp. 527–531, 2005.
49. F. Mormann, T. Kreuz, R. G. Andrzejak, P. David, K. Lehnertz and C. E. Elger, "Epileptic seizures are preceded by a decrease in synchronization," *Epilepsy Res.*, vol. 53, pp. 173–185, 2003.
50. C. Hugenii, *Horoloquium Oscilatorum*, Paris, France, 1673.

51. M. G. Rosenblum, A. S. Pikovsky, J. Kurths, "Phase synchronization of chaotic oscillators," *Phys. Rev. Lett.*, vol. 76, pp. 1804–1807, 1996.
52. D. Gabor, Theory of communication, *Proceedings of IEE London* 93, pp. 429–457, 1946.
53. P. Panter, *Modulation, Noise, and Spectral Analysis*, McGraw-Hill, New York, 1965.
54. S. Sabesan, N. Chakravarthy, L. Good, K. Tsakalis, P. M. Pardalos and L. D. Iasemidis, "Global optimization and spatial synchronization changes prior to epileptic seizures", University of Coimbra, Workshop on *Optimization in Medicine*, Coimbra, Portugal, July 20–22, 2005, Springer Verlag, pp. 103–125, 2007.
55. N. H. Packard, J. P. Crutchfield, J. D. Farmer and R. S. Shaw, "Geometry from time series," *Phys. Rev. Lett.*, vol. 45, pp. 712–716, 1980.
56. F. Takens, "Detecting strange attractors in turbulence," in *Dynamical Systems and Turbulence, Lecture Notes in Mathematics*, D. A. Rand and L. S. Young, Eds., Heidelberg: Springer-Verlag, 1981.
57. P. Grassberger and I. Procaccia, "Measuring the strangeness of strange attractors," *Physica D*, vol. 9, pp. 189–208, 1983.
58. P. Grassberger and I. Procaccia, "Characterization of strange attractors," *Phys. Rev. Lett.*, vol. 50, pp. 346–349, 1983.
59. E. J. Kostelich, "Problems in estimating dynamics from data," *Physica D*, vol. 58, pp. 138–152, 1992.
60. H. D. I. Abarbanel, *Analysis of Observed Chaotic Data*, New York: Springer Verlag, 1996.
61. H. Haken, *Principles of Brain Functioning: A Synergetic Approach to Brain Activity, Behavior and Cognition*, Springer-Verlag, Berlin, 1996.
62. J. A. Vastano and E. J. Kostelich, "Comparison of algorithms for determining Lyapunov exponents from experimental data," in *Dimensions and Entropies in Chaotic Systems: Quantification of Complex Behavior*, G. Mayer-Kress, Ed., Berlin: Springer-Verlag, 1986.
63. L. D. Iasemidis, D. S. Shiau, W. Chaovalitwongse, J. C. Sackellares, P. M. Pardalos, P. R. Carney, A. Prasad, B. Veeramani, and K. Tsakalis, "Adaptive epileptic seizure prediction system", *IEEE Trans. Biomed. Eng.*, vol. 50(5), pp. 616–627, 2003.
64. L. D. Iasemidis, D.-S. Shiau, P. M. Pardalos, W. Chaovalitwongse, K. Narayanan, A. Prasad, K. Tsakalis, P. Carney and J. C. Sackellares, "Long-term prospective on-line real-time seizure prediction", *J. Clin. Neurophysiol.*, vol. 116, pp. 532–544, 2005.
65. W. Chaovalitwongse, L. D. Iasemidis, P. M. Pardalos, P. R. Carney, D.-S. Shiau and J. C. Sackellares, "Performance of a seizure warning algorithm based on nonlinear dynamics of the intracranial EEG", *Epilepsy Res.*, vol. 64, pp. 93–113, 2005.
66. C. Domb, in *Phase Transitions and Critical Phenomena*, C. Domb and M. S. Green, Eds., New York: Academic Press, 1974.
67. R. Esteller et al., "Continuous energy variation during the seizure cycle: Towards an on-line accumulated energy" *Clinical Neurophysiology* 116, pp. 517–526, 2005.
68. M. G. Rosenblum, A. S. Pikovsky, J. Kurths, "From phase to lag synchronization in coupled chaotic oscillators," *Phys. Rev. Lett.*, vol. 78, pp. 4193–4196, 1997
69. V. S. Afraimovich, N. N. Verichev, M. I. Rabinovich, "General synchronization," *Izv VysshUch Zav Radiofizika*, vol. 29, pp. 795–803, 1986.
70. N. F. Rulkov, M. M. Sushchik, L. S. Tsimring, H. D. I. Ababarnel, "Generalized synchronization of chaos in directionally coupled chaotic systems," *Phys. Rev. E*, vol. 51, pp. 980–994, 1996.
71. H. Fujisaka and T. Yamada, "Stability theory of synchronized motion in coupled-oscillator systems," *Progr. Theor. Phys.*, vol. 69, pp. 32–37, 1983.
72. L. D. Iasemidis and J. C. Sackellares, "Chaos theory in epilepsy," *The Neuroscientist*, vol. 2, pp. 118–126, 1996.
73. J. C. Sackellares, L. D. Iasemidis, R. L. Gilmore and S. N. Roper, "Epileptic seizures as neural resetting mechanisms", *Epilepsia*, vol. 38(S3), p. 189, 1997.
74. D. S. Shiau, Q. Luo, R. L. Gilmore, S. N. Roper, P. Pardalos, J. C. Sackellares and L. D. Iasemidis, "Epileptic seizures resetting revisited", *Epilepsia*, vol. 41(S7), p. 208, 2000.

75. W. A. Chaovalitwongse, L. D. Iasemidis, P. M. Pardalos, P. R. Carney, D.-S. Shiau and J. C. Sackellares, "Reply to comments by F. Morman, CE Elger, and K. Lehnertz on the performance of a seizure warning algorithm based on the dynamics of intracranial EEG", *Epilepsy Res.*, vol. 72, pp. 85–87, 2006.
76. W. A. Chaovalitwongse, L. D. Iasemidis, P. M. Pardalos, P. R. Carney, D.-S. Shiau and J. C. Sackellares, "Reply to comments by M. Winterhalder, B. Schelter, A. Achulze-Bonhage and J. Timmer on the performance of a seizure warning algorithm based on the dynamics of intracranial EEG", *Epilepsy Res.*, vol. 72, pp. 82–84, 2006.
77. J. C. Sackellares, D.-S. Shiau, J. C. Principe, M. C. K. Young, L. K. Dance, W. Suharitdamrong, W. Chaovalitwongse, P. M. Pardalos and L. D. Iasemidis, "Predictability analysis for an automated seizure prediction algorithm", *J. Clinical Neurophysiol.*, vol. 23, pp. 509–520, 2006.
78. P. Suffczynski, S. Kalitzin and F. H. L. Da Silva, "Dynamics of non-convulsive epileptic phenomena modeled by a bistable neuronal network", *Neuroscience*, vol. 126, pp. 467–484, 2004
79. D. Liley, P. Cadusch and M. Dafilis, "A spatially continuous mean field theory of electrocortical activity", *Network: Comp. in Neural. Sys.*, vol. 13, pp. 67–113, 2002.
80. E. Grassi, K. S. Tsakalis, S. Dash, S. V. Gaikwad, W. MacArthur and G. Stein, "Integrated system identification and PID controller tuning by frequency loop-shaping", *IEEE Trans. Control Sys. Tech.*, vol. 9, pp. 285–294, 2001.
81. K. J. Astrom and L. Rundqwist, "Integrator windup and how to avoid it", *Proceedings of the 1989 American Control Conference*, vol. 21–23, pp. 1693–1698, June 1989.

An Expressive Body Language Underlies *Drosophila* Courtship Behavior

Ruedi Stoop and Benjamin I. Arthur

Abstract Significant progress in the understanding of the neuroanatomical and chemical basis of *Drosophila* courtship behavior has recently been obtained by means of genetic techniques. Progress in understanding the changes in the courtship behavior in response to the genetically modified central nervous system was, however, hampered by the lack of an appropriate methodology for the measurement of behavior. Here, we propose an operational definition of behavior that provides a sensitive quantification of the differences among *Drosophila* courtship behaviors. Using this approach, we gain evidence that a highly specific and expressive body language underlies *Drosophila* courtship, offering the largest individual expression bandwidth for normal males courting mature females. We also find that normal male *Drosophila* perform a switch from male to female behavior if they are together with males carrying the fruitless gene. From this we conjecture that the courtship behavior of normal *Drosophila* is based upon a sensor-driven bistable neurodynamical system, whereas the fruitless mutations lack this adaptive ability. We anticipate our assay and the developed methodological approach to be a starting point for systematic investigations of the relationship between genes and behavior.

Keywords Markov model · Symbolic dynamics · Behavior · Courtship Language · Time series

1 Introduction

Courtship of the fruit fly *Drosophila* is a standard example of genetically hardwired behaviors. Whereas normal male *Drosophila melanogaster* court females only, using advanced genetic techniques applied to the *fru* gene [1, 2] it is now possible to

---

R. Stoop (✉)
Institute of Neuroinformatics, Physics Department, University/ETH Zürich,
Winterthurerstr. 190, 8057, Zürich, Switzerland
e-mail: ruedi@ini.phys.ethz.ch

B.I. Arthur designed the experiments collected and coded the data, R. Stoop designed and performed the data processing and wrote the paper.

generate males that only court males or court both genders, and females that court males or only court females. This provides hard evidence that gene information in addition to how living beings are built also defines to a considerable extent how they behave [3, 4]. Why animals court, is largely *terra incognita*. Since *Drosophila* is under strong evolutionary pressure, pre-copulation courtship could serve as an information platform for assessing a potential partner's suitability for passing on genes. In order to investigate how this purpose could be achieved within the genetical constraints, single normal females in the immature, mature, and mated states were paired with single normal males (resulting in 6 groups of protagonists) and

Table 1 Encoding: 37 fundamental acts are encoded into numbers, some of which are sex-specific

Drosophila acts:		
1	abdobend	female
2	abdotwist	female
3	attemptcop	male
4	circling	male
5	copulation	male
6	decamp	male/female
7	fencing	male
8	following	male
9	grooming forlegs	male/female
10	grooming hindlegs	male/female
11	headpos	male/female
12	kick hindlegs	female
13	licking	male
14	orientation	male
15	ovipext	female
16	run	male/female
17	standing	male/female
18	still	male/female
19	tapping	male
20	walk left	male/female
21	walk right	male/female
22	wingext left	male
23	wingext	right male
24	wingflicks left	male/female
25	wingflicks right	male/female
26	wingflutter	female
27	wingspread	male/female
28	wingwave	male
29	wingflicks unspec.	male/female
30	grooming midlegs	male/female
31	tapping forelegs	male
32	kick midlegs	female
33	walk unspec.	male/female
34	kick unspec.	male/female
35	wingflap	male/female
36	run right	male/female
37	run left	male/female

fruitless mutant males [5] were paired with either mature females or with mature normal males (4 more groups of protagonists). The emergent courtship behavior has tactile, gustatory, olfactory, acoustic and visual sensory dimensions [6, 7]. In pre-copulation courtship of *D. melanogaster*, visual information is salient [8, 9] and easy to access, which is why we focus on this component. This sensory reduction is not problematic, as one may faithfully reconstruct the multi-dimensional system from one single component [10]. Although our analysis is based on time series of one protagonist at a time, we also obtain insight into the interplay between partners. We performed the visual recordings at the almost neuronal resolution of 30 frames per second. By basing our analysis on the 37 independent (*"fundamental"*) behavioral acts that, on this time-scale, compose the courtship behavior (see Table 1), we eliminated potential ambiguities in the definition of the courtship states. For the recordings, using frame-by-frame inspection and focusing on one single protagonist, the starting points of the fundamental acts were detected. The corresponding indices (see Table 2) were written into files on which our further analysis is based. For each possible experimental constellation, five trials of the experiment were performed, using different individuals. The individuals were not previously screened for how efficiently they would court.

To estimate the courtship information content of the files, we evaluated the Shannon entropies $h_s = -\sum_{i=1}^{n_s} p_i \log p_i$, where n_s is the number of symbols used, i indexes the fundamental acts (or their corresponding symbols) and p_i their probability of occurrence [10]. For similar behavior, we expect to measure similar values of h_s. The averaged results shown in Fig. 1 support all basic observations known to hold for *Drosophila* courtship: Overall, male *Drosophila* use a richer repertoire (full vs. dotted line). Males use a richer repertoire to court an already mated female (experiment 1 vs. experiment 4). In experiment 5 (mature female, *fruitless* male), a decreased activity is accompanied by a reduced repertoire by both protagonists. When normal males are paired with *fruitless* mutant males (experiment 3), the entropies support the visual observation that under these conditions the normal males adopt a female-like role and that the *fruitless* males express a very pure male-like role. Otherwise, a crossing of the lines would result.

Table 2 Correspondence between matrix indices and *Drosophila* behavioral vectors. Boldface letters identify the observed protagonists

1	≙	fruitless males-mature females
2	≙	mature **F**emales-fruitless males
3	≙	fruitless males-normal males
4	≙	normal **M**ales-fruitless males
5	≙	normal **M**ales-mated females
6	≙	mated **F**emales-normal males
7	≙	normal **M**ales-mature females
8	≙	mature **F**emales-normal males
9	≙	normal **M**ales-immature females
10	≙	immature **F**emales-normal males

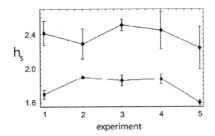

Fig. 1 Shannon entropies h_s calculated for the 5 courtship experiments. Each data point corresponds to the average obtained from 5 different *Drosophila* individuals. Protagonists (bottom/top) (1) mature females/normal males; (2) immature females/normal males; (3) normal males/*fruitless* males; (4) mated females/normal males; (5) mature females/*fruitless* males. In (experiment 3), the value obtained for normal males has been connected with the female data points, and that of *fruitless* males with those of the normal males (error bars: standard deviation)

2 Scrutinizing the Notion of Behavior

Taken as a measure of behavior, the Shannon entropy has profound deficits. h_s depends strongly on the number of the symbols used, which may be partially responsible for the overall difference between female and male behavior (more symbols are accessible to males than to females, see Table 2). As a consequence, an unbiased comparison of information is restricted to within the female group (i.e. immature, mature, mated), or the male group (i.e. normal and *fruitless* males). Even more importantly, the Shannon entropy ignores the individuality of the symbols and the order in which they appear, which both are obvious important characteristics of behavior. Entirely different temporal sequences, composed of different dominant symbols, may therefore lead to identical values of h_s. The fact that h_s is maximal for an equi-distributed random process (that would correspond to a fortuitous testing of all available symbols), is a further indicator of the difficulty of establishing a direct relationship between h_s and behavior.

If during pre-copulation courtship a potential partner's suitability for passing on genes is assessed [6, 11, 12], we expect to observe a large behavioral variability within genetically normal animals and a reduced variability within the *fruitless* mutants. In addition, we expect that behavioral classes emerge that distinguish between normal and genetically mutant *Drosophilae*, between the different sexes, and possibly also between their different stages of sexual development. For this analysis, the traditional tools *courtship index* (the fraction of the whole time spent by an animal in obvious courtship) and the *mating success* (the fraction of successful versus total attempts of copulation), are unsuited. They are both integral measures [3, 6, 11] which cannot resolve more specific aspects of behavior. Precursors in our attempt to refine the description of courtship are behavioral transition graphs [8, 13–15]. One would expect to see genetic variations reflected in topological and metric modifications of these graphs [8, 13], as well as in changed salient individual behaviors. Unfortunately, the usual Markov transition graph approach allows only few behavioral states to be considered. By concatenating several fine acts of behavior into macroscopic (i.e., temporally extended) states, we would lose the ability to address fine-scale issues. Moreover, first-order Markov models only take one transition step into account. As a consequence, no precise information about longer *successions* of fundamental acts, which seem essential for behavior, can be extracted. Higher-order models could be converted into first order, but only at the cost of exponential growth of the model in the order, which quickly leads beyond tractability. Moreover, comparability of the resulting graphs would be restricted to within each male/female class, because some of the fundamental acts of Table 2 only apply to females, others to males only.

To remedy this, we worked out an alternative way of characterizing behavior. Human courtship proceeds along stereotyped tracks: Take a girl for a coffee, go with her to the movies, walk her home (often several times), get invited for a last drink, etc. Here, we posit that for *Drosophila* a similar characterization applies. It is conceivable that precisely *sequences* of fundamental acts provide a powerful description of behavior [16], Similarly to the role of letters/words with respect to words in spoken language. Behavior is usually defined as *"the aggregate of actions or reactions of an object or organism, usually in relation to the environment"* [17]. We propose to characterize behavior by its *set of irreducible closed orbits of fundamental acts* (indecomposable orbits of no further dissectable acts of behavior). As redundancy is an important element of biological information exchange, we may expect the characteristic behaviors to occur repeatedly, embedded within other activities, and assume that the characteristic behaviors are robust enough to statistically stand out from this random environment. This definition of behavior leads directly to the following computational implementation: From the experimental data, we extract the set of irreducible closed orbits s_i, $i = 1\ldots n$. A template vector of dimension n is then formed by assigning to each of these orbits a particular vector component. As a function of their length l, the number of closed orbits can be expected to grow no more than $N(l) \simeq e^{lh_{\text{top}}}$ where h_{top} is the topological entropy of the process [10]. l has to be chosen taking into account the size of the data and the behavioral context. In our case, we went up to a length of $l = 7$ symbols, for which we obtained 181 irreducible closed orbits. The template vector can be interpreted as defining *Drosophila*'s behavioral space with respect to all of the experiments considered. For each protagonist j, we constructed a behavioral vector $\hat{b}_j$, by filling the entries of the template vector with the numbers of occurrences of the corresponding orbits. Courtship behavior is not a continuous, but a highly variable process that occurs in bouts that can be seen as a prototype of complex behavior [18]. The description of behavior in terms of closed sequences of fundamental acts is thus closely related to the dissection of complex chaotic motion into periodic orbits [19], which provides a variable-order Markov approximation to the underlying system. Typically, courtship proceeds in a transient fashion, where only after completion of a certain context (possibly on first attempt), one is allowed to access the next context.

In order to obtain a measure for behavioral similarity/dissimilarity, we use normalized behavioral vectors b_i, and the projectional properties of the scalar product. By projecting onto the orbits common to both compared behaviors, the scalar product assesses their similarity in a simple way. The influence of a multiple occurrence of orbits can be monitored by weighting the behavioral vector componentwise by an exponent b ∈ $\Re$. This method is borrowed from statistical physics, where β's role is that of an inverse temperature (e.g., [10]). In this way, we characterize the behavioral similarity between two protagonists by a one-parameter family

$$m_{i,j}^\beta = \langle b_i, b_j \rangle^\beta, \; \beta \in \Re, \tag{1}$$

where $\langle .,. \rangle$ denotes the scalar product between the behavioral vectors of the protagonists i and j, respectively. All pairwise similarities can be collected in a matrix $M(\beta) = (m^\beta)_{i,j}$. For $\beta \geq 0$, large/small matrix entries indicate similar/dissimilar behaviors. For $\beta = 1$, the natural measure is obtained (i.e. the measure that is based on precisely how often cycles appear), whereas for $\beta = 0$, this information is reduced to the topological aspect (only the occurrence or absence of a particular cycle matters [10]). Bearing in mind that aberrant behavior of one single fly could entirely dominate the average behavior, we will concentrate on the topological characterization. Note that our notion of similarity is only an approximate one that does not include transitivity.

3 Individual vs. Class Behavior

Once the individual protagonists are listed according to their experimental group (see Table 1 for the correspondence ("≙") between the matrix indices and the experiments), the similarities emerge as shown in Fig. 2a. As expected, the results display a large individual variability among protagonists of the same class. Note that identical patterns imply that identical behavioral sequences were used, but neither that they were displayed in the same order nor (for $\beta = 0$) that they were displayed with the same frequency. To assess the amount of bias introduced by a pooling from individual to class behavior, we performed a two-sided mean-difference significance test for the hypothesis that thesets of similarities obtained for two protagonists would originate from the same distribution. Although we expect this to underestimate the behavioral discriminability, the original class structures are strongly conserved (Fig. 2b). Moreover, from the data, the identification of individuals sharing, or showing unusual, behavior within one class, is straightforward.

As a general feature, individual females – in particular mature females in the presence of normal males – use only few behavioral alternatives. In contrast, the individuality of males in the presence of mature females appears to be of central importance. This we take as an indication that it is the individual male response to the more stereotypical female behavioral patterns that defines the salient information flow in *Drosophila* courtship. At a given confidence level, majority voting can be used to estimate whether a class could stand out against another class. At a confidence level of $p = 0.85$, all classes are distinct. At $p = 0.95$, experiments

An Expressive Body Language Underlies *Drosophila* Courtship Behavior 221

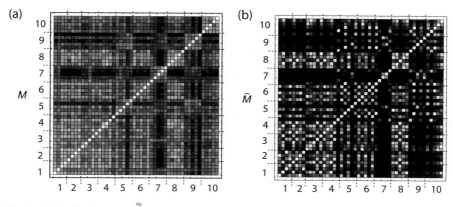

Fig. 2 (a) Similarity matrix $\widetilde{\widetilde{M}}(\beta = 0)$ evaluated for all individuals (light shading: high similarity). Strongly individual normal male behavior under "normal conditions" is contrasted by a lack of individual behavior in all experiments involving fruitless mutations. Emerging patterns correlate largely with group boundaries indicated by *dashed lines*. (b) Behavioral discriminability, corroborating individual variability on top of group coherence (mean-difference test, *light shading*: low discriminability, except for the diagonal)

involving the fruitless mutation, female behavior and male behavior are identifiable. At confidence level $p = 0.99$, the normal male behavior towards mature, and towards immature females can be distinguished from the rest. Above this confidence, only the male behavior towards mature females stands out. Although these tests underestimate considerably the behavioral discriminability, they corroborate our approach's ability to discriminate in a quantitative and fine-grained manner among the different behaviors involved in *Drosophila* courtship. We also conclude that pooling of individual into classes of behaviors adds only a tolerable amount of *a priori* information. The actual pooling was performed by adding all behavioral vectors of a class and then renormalizing the resulting vector. The behavioral similarity among the ten different classes is thus captured in the matrix

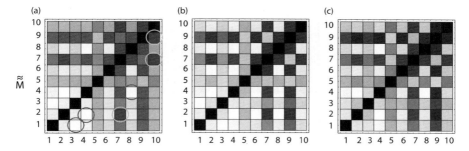

Fig. 3 Class similarities based on (**a**) the closed orbit analysis of the original data, (**b**) the closed orbit analysis of surrogate data, (**c**) on the symbol probabilities only (see text). Density plots over $[0; \text{maxcorr}] = [0; 1]$). *Lighter shading* indicates higher similarities, diagonal elements were set to 0. *Red* and *blue circles* indicate maximal and minimal similarities, respectively

$$\widetilde{\widetilde{M}}(\beta)_{i,j} = \langle \hat{b}_i, \hat{b}_j \rangle^\beta \qquad (2)$$

where $\hat{b}_i$, $\hat{b}_j$, $i, j \in \{1, \ldots, 10\}$ are the pooled behavioral class vectors. A density plot of $\widetilde{\widetilde{M}}(\beta = 0)$ as obtained from the experimental data is shown in Fig. 3a.

4 Evidence of a Nontrivial Grammar

We then asked to what extent the closed orbits could contribute to an identification of the experimental classes. In the first comparison, we characterized each protagonist by a behavioral vector the entries of which were the natural symbols probabilities of the individual files (i.e., vectors of length 37). It is remarkable that on this level of simplification the similarities/dissimilarities between the classes are already expressed (see Fig. 3a/c). The characteristic stripes and peaks emerge at the same places as for the orbit analysis. This implies that the distinction among classes is to a large extent rooted in symbol probabilities. Could *Drosophila* courtship communication be solely based upon symbol probabilities? In a second experiment, we repeated the closed orbit analysis on shuffled data files (surrogate data method).

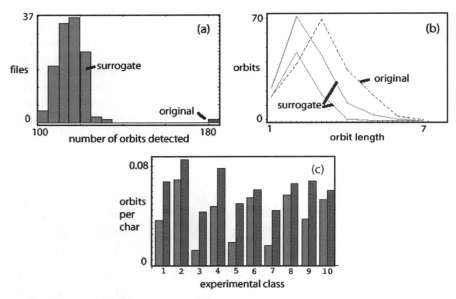

Fig. 4 Closed orbits found in the original and in the surrogate files. (**a**) Histogram of the number of (different) closed orbits found in 120 surrogate runs, where each file was shuffled individually. (**b**) Maximal/minimal number of orbits of a given length found across 120 surrogate runs. (**c**) Number of closed orbits per character for the different experimental classes. Surrogate files (*red*) contain substantially fewer orbits than the original files (*blue*). Roughly one third of the surrogate orbits are not present in the original files

Whereas the original closed orbits analysis provides a variable-order Markov model approximation of the original data, here the underlying Markov model is of zeroth order. Of all the different models that realize a given symbol distribution, the surrogate method returns the most probable ones. The results obtained for this experiment demonstrate that the basic (dis)similarities between the classes are also maintained in this approximation (see Fig. 3b). A detailed comparison of the three similarity matrices displayed in Fig. 3 confirms, however, that the closed orbits not only serve as the substrate for the communication among individuals, but also enhance the distinguishability between classes. To see this, note that the submatrices composed of non-fruitless vs. fruitless experiments are most clearly worked out in the original data, a feature that is corroborated by a network analysis of the symbol transition networks.

Moreover, the detailed analysis of surrogate data yields that significantly less and different closed orbits are detected if compared to the original data (see Fig. 4). This implies that a large portion of the orbits found in the original data are *a priori* unlikely ones. Therefore, an additional, nonaccidental, structure, which can be interpreted as a grammar, underlies *Drosophila* courtship behavior.

5 Extremal Class Similarities

General observations from the similarity matrix $\widetilde{M}$ of the original data (Fig. 3a) are that male *Drosophila* vary their behavior strongly with their partner class (no peaks of similarity within the columns indexed by 5, 7, 9), and that females vary their behaviors to a much lesser extent. For example, the behaviors of mature and mated females are rather similar, but dissimilar from the immature. This culminates in the marked distinction of the normal male behavior towards mature and immature females (columns 7 and 9), against all other behaviors. The largest contrast is obtained between columns 7 and 8, which highlights the special role of courtship between mature females and males. Dark fields adjacent to the diagonal indicate the large behavioral distance between male/female protagonists from the same experiment. *Maxima of similarity* are obtained for the following pairs of classes: (4, 2): normal Males behave towards *fruitless* males similar to mature Females towards *fruitless* males (0.90); (8, 4): mature Females behave towards normal males similar to normal Males towards *fruitless* males (0.88); (3, 1): fruitless males behave towards normal males similar to fruitless males towards mature females (0.87). Whereas female *Drosophila* do not discriminate particularly between *fruitless* and normal males, when in contact with *fruitless* males, normal males switch to female behavior. The high similarities displayed at the lower left corner of $\widetilde{M}$ (indices [1,...,4]) express the lack of behavior adaptation by, or caused by, the *fruitless* mutants. The *lowest similarities* are obtained for (10, 7): immature Females behave towards normal males nearly orthogonally to normal Males towards mature females (0.22); (10, 9): immature Females behave towards normal males nearly orthogonally to normal Males towards immature females (0.23); (7, 2): normal Males behave towards mature females nearly orthogonally to mature Females towards *fruitless* males (0.25). Clearly, the minima express the large behavioral distance between male and female *Drosophila* behaviors.

6 Conclusions

Our concept of behavioral similarity focuses on closed sequences of fundamental acts instead of long-time averages or isolated behavioral acts. This allows for a detailed quantitative comparison of behavior, which we have used to obtain a refined insight into the similarity/dissimilarity of individual, and classes of, behaviors in *Drosophila* pre-copulatory courtship. The performed analysis provides hard evidence that in this refined description, group behavior can be identified in agreement with more crude measures of behavior and with generally accepted observations by experimenters. As an example, our method objectively confirms earlier observations [20] that the behavior of females mainly depends on their reproductive status (immature/mature/mated). A detailed analysis using surrogate data uncovers the presence of a nontrivial grammar in *Drosophila* courtship that offers a large bandwidth for pre-copulatory communication. Individual males in particular, widely use this expression space. This indicates that during courtship more than just group membership is exchanged. Potentially transferred information range from encoded details on the sender's genetic configuration to situation-dependent decisions (as the "better mate vs. no mate" trade-off). Our method is general enough to be applicable to a wide variety of comparative behavioral, or behavioral neurogenetics, studies that critically depend upon a sufficiently detailed quantitative analysis. The result that normal males in the presence of *fruitless* males display a behavior similar to that of females being courted, is of interest. It implies that normal males perform a context-triggered male-to-female behavioral switch. In the behavior by females and by the *fruitless* males, such a contextual switch is unobserved. That the behavioral system in *fruitless* males is severely handicapped is corroborated by the fact that they also court each other [21]. *D. melanogaster* courtship behavior thus has several facets: A genetically hardwired one that endures for a lifetime, but can be genetically manipulated [22]. In addition, there is learned behavior, generally lasting from a few hours to days [9]. In contrast to these behavioral paradigms, the newly discovered behavioral phenomenon involves a context-dependent switching from the male to the female behavioral role. Whether this phenomenon is just a relic from evolution, or whether it plays an active evolutionary role, remains open to investigation. Since the male and female neural systems appear to be based on identical neuronal circuitries [3, 4], it is conceivable that sexual behavioral orientation in male *Drosophila* is implemented in the form of a bistable neuro-dynamical system that possesses two major basins of attraction. By means of sensory (e.g., olfactory) contextual input, the male behavior is reversibly pushed towards male or female behavior, whereas the females' sensory input would not offer this possibility. This view, which challenges the picture of structurally different male/female neuronal systems, would be in agreement with the major findings of Ref. [3]. Observing the neuronal correlates that correspond to the salient behavioral courtship elements, will be an efficient method for studying how courtship behavior is implemented on the neuronal circuitry level.

7 Methods

7.1 Fly Stocks

The *D. melanogaster* flies used in this study are of the wild-type Canton S (CS) strain and mutants carrying an allele of the *fruitless* mutation (fru^1) [1, 2]. Raised on a corn-meal/agar/molasses/yeast medium at 25°C on a 12:12 L: D cycle, the flies were sexed and the naive males and virgin females were isolated within two hours after eclosion under cold anesthesia (4°C). Until recording, the males were kept individually in test tubes, where mature male and female flies used for the recording were 4–5 days old. The mutant males used in the assays were homozygous for fru^1.

7.2 Data Acquisition

A pair of flies was transferred, by aspiration, into a cylindrical mating chamber of dimensions 0.8 cm diameter and 0.5 cm height. Environmental conditions were fixed at 25°C and 75% humidity. A 5-min episode (or until copulation in the case of the mature virgins) was recorded with a Sony Hi8 video/audio camera, using 29.97 frames per second. The videotape recordings were converted from Hi8 to digital video, and then compressed and converted into MPEG 1 format, using Cleaner 6 software (Discreet, New York, NY). The files were loaded onto the THEME coder (Patternvision Inc., Iceland) and analyzed frame by frame. The beginning and ending of every act was systematically registered, for both protagonists (resulting in two files, one for each protagonist). The data were analyzed frame by frame (30 frames per second) using an interactive multimedia program. The beginning and ending of every behavior was systematically registered for the male and female of each pair to a 1/30 of a second resolution. At this time resolution, we discovered novel behavior elements that were invisible to the unaided eye [Arthur et al., in prep].

7.3 Fundamental Behavioral Acts

The fundamental behavioral acts coded for were the following:

Female behavior elements: *Abdotwist*: a twisting of the abdomen sideways and down-wards. *Kicking*: applying tarsi forcefully against partner. *Ovipositor extrusion*: telescoping extension of ovipositor. *Wiggle*: Wagging of abdomen and wings in opposite directions. *Wingflutter*: left and right flicks in succession.

Male behavior elements: *Orientation*: Fly is facing the partner, and may involve little motion such as turning to keep facing the partner, leaning over or no motion. *Following*: locomotion in directed pursuit of a partner. *Tapping*: touching partner with tarsi of prothoracic forelegs. *Fencing*: face-to-face and engaged in sparring with prothoracic forelegs. *Wing extension*: wing stretched out away (i.e. perpendicular to the head abdomen axis) from the body biased to the left or right. *Circling*: a sideways skid along a semi-circular path around a partner. *Licking*: proboscis contact with partner. *Abdominal bending (Abdobend)*: abdominal curling under thorax towards the head direction. *Attempted copulation*: abdominal curling directed towards a partner. *Copulation*: sustained genital connection between male and female.

Common behavior elements: *Abdominal vibration*: vertical up and down movement of the abdomen. *Decamp*: an evasive jump away from partner, usually involving a somersault. *Grooming*: rubbing of tarsi together, forelegs, midlegs or hindlegs. *Standing*: absence of locomotion with no indication of activity directed towards the partner. *Still*: no detectable movement of any body parts during standing. *Walking (walk)*: locomotion that is not directed towards courting the partner. *Wing Flicks*: brief spasmic movement of wings to and from the antero-posterior axis of the body in rapic succession. *WingScissoring*: both wings moved away from and back to the body in a rapid scissor-like manner. *Wing Spread*: Both wings are stretched out away from the body to give a wide V-shape forming an angle of at least 90 degrees that is bisected by the anterior to posterior axis of the fly. We set a criterion that any continuous action that had a beginning and ending that was clearly observable at the temporal resolution of 30 frames per second fit our definition of behavior element. The high spatial resolution and sub-second temporal resolution enabled us to observe novel fast action behavior elements. Not all of the described elements were found in the reported data set.

7.4 Closed Orbits Extraction

From the data, for simplicity we only used the symbols indicating the beginning of an act. Starting with the first symbol of the file, we scan through the file until a first repetition of the symbol is found. This then is repeated for the following symbols. The procedure is organized such that automatically only irreducible (indecomposable) orbits are collected.

7.5 Surrogate Data Analysis

For the surrogate data, we shuffled the data within individual files by means of permu-tations. Note that in this way, there is an increased tendency for periodic orbits to occur (since successive repetitions of some of the symbols are very unlikely) that does not show up in the original files. The displayed results were not corrected for this effect. A corresponding correction preserves the reported findings.

References

1. J. Hall, *Science* **264**, 1702–1714 (1994).
2. M. Adams et al., *Science* **24**, 2185–2195 (2000).
3. P. Stockinger, D. Kvitsiani, S. Rotkopf, L. Tirian, B.J. Dickson, *Neural circuitry that governs Drosophila male courtship behavior*, Cell **121**, 795–807 (2005).
4. E. Demir, B.J. Dickson, *Fruitless splicing specifies male courtship behavior in Drosophila*, Cell **121**, 785–794 (2005).
5. D.A. Gailey, J.C. Hall, *Behavior and cytogenetics of fruitless in Drosophila melanogaster: different courtship defects caused by separate, closely linked lesions*, Genetics **121**, 773–785 (1989).
6. H.T. Spieth, *Courtship behavior in D.*, Ann. Rev. Entomol. **19**, 385–405 (1974).
7. L. Giarratani, L.B. Vosshall, *Toward a molecular description of pheromone perception*, Neuron **39**, 881–883 (2003).
8. T.A. Markow, *Behavioral and sensory basis of courtship*, Proc. Natl. Acad. Sci. USA **84**, 6200–6204 (1987).
9. H.V. Hirsch, L. Tompkins, *The flexible fly: experience-dependent development of complex behaviors in D.*, J. Exp. Biol. **195**, 1–18 (1994).
10. J. Peinke, J. Parisi, O.E. Roessler, R. Stoop, "Encounter with Chaos", Springer, Berlin (1992).
11. H.B. Dowse, J.M. Ringo, K.M. Barton, *A model describing the kinetics of mating in Drosophila*, J. Theor. Biol. **121**, 173–183 (1986).
12. P. Welbergen, B.M. Spruit, F.R. van Dijken, *Mating speed and the interplay between female- and male courtship responses in Drosophila melanogaster (Diptera: Drosophilidae)*, J. Insect Behav. **5**, 229–244 (1992).
13. T.A. Markow, S.J. Hanson, *Multivariate analysis of Drosophila courtship*, Proc. Natl. Acad. Sci. USA **78**, 430–434 (1981).
14. A. Hoikkala, S. Crossley, *Copulatory courtship in Drosophila: Behavior and songs of D. birchii and D. serrata*, J. Insect Behav. **13**, 71–86 (2000).
15. A. Hoikkala, S. Crossley, C. Castillo-Melenedez, *Copulatory courtship in Drosophila birchii and D. serrata, species recognition and sexual selection*, J. Insect Behav. **13**, 361–373 (2000).
16. S. Chen, A.Y. Lee, N.M. Bowens, R. Huber, E.A. Kravitz, *Fighting fruit flies: A model system for the study of aggression*, Proc. Natl. Acad. Sci. USA **99**, 5664–5668 (2002).
17. Wikipedia, http://en.wikipedia.org/wiki/Behavior.
18. R. Stoop, N. Stoop, L.A. Bunimovich, *Complexity of dynamics as variability of predictability*, J. Stat. Phys. **14**, 1127–1137 (2004).
19. R. Stoop, C. Wagner, *Scaling properties of simple limiter control*, Phys. Rev. Lett. **90**, 154101 (2003).
20. R.J. Greenspan, J.F. Ferveur, *Courtship in D.*, Ann. Rev. Genet. **34**, 205–232 (2000).
21. A. Villella, D. A. Gailey, B. Berwald, S. Oshima, P.T. Barnes, J.C. Hall, *Extended reproductive roles of the fruitless gene in Drosophila melanogaster revealed by behavioral analysis of new fru mutants*, Genetics **147**, 1107–1130 (1997).
22. B.I. Arthur, J.-M. Jallon, B. Caflisch, Y. Choffat, R. Noethiger, *Sexual behavior in D. is irreversibly programmed during a critical period*, Curr. Biol. **8**, 187–190 (1998).

Speech Rhythms in Children Learning Two Languages

T. Padma Subhadra, Tanusree Das and Nandini Chatterjee Singh

Abstract In an increasingly global world where large populations of children acquire two languages there is very little information on when bilingual children exhibit speech rhythms that are language-specific. We examine the rhythmic features of speech from 70 children between 5 and 8 years, learning two languages, English and Hindi and 11 adults who are fluent speakers of both languages. We relate variability in syllable duration to speech rhythm and find that adults exhibit significant differences in durational variability between the two languages.

We estimate syllable durations and calculate durational variability for both languages and find that at 5 years children exhibit similar durational variability for both English and Hindi. However around 7 years of age we find that durational variability for English becomes significantly larger than that of Hindi. Our findings are in accordance with the rhythmic classification of Hindi and English as syllable- and stress-timed languages respectively wherein durational variability is greater in stress-timed languages (English) than in syllable-timed languages (Hindi). We therefore suggest that children learning two languages exhibit characteristic speech rhythms around 7 years of age.

Keywords Language rhythm · Hindi · English · Children · Syllable duration · Bilingual

1 Introduction

Theories on language acquisition must provide patterns of language development not only in monolingual speakers, but also in children who are acquiring multiple languages. In an increasingly global world, a multilingual capability offers competitive advantages and there is evidence [1] to indicate that the number of children who are acquiring a second language has steadily increased in the last few years. While a few studies have been directed towards studying the acquisition of intonation [2–4]

N.C. Singh (✉)
National Brain Research Centre, Nainwal Mode, NH – 8, Manesar, 122 050, Haryana, India
e-mail: nandini@nbrc.ac.in

in spoken language, the acquisition of speech rhythm in speech particularly in children learning more than one language has hardly received any attention. In the present paper, we study the acquisition of speech rhythm in children learning two languages, in this case Hindi and English. The study is motivated by two factors – in the present study, which is from India, a large population of children and adults is bilingual. The language of learning and instruction in many schools is English and children grow up learning at least one Indian language and English. Interestingly, based on the rhythm class hypothesis, English has been classified as stress-timed and Hindi has been classified as syllable-timed. This is therefore a unique population to study patterns of development in the acquisition of multiple language rhythms.

Rhythm as a language-specific linguistic feature has to be acquired by children learning language and it has been suggested that speech rhythm is one of the earliest aspects of speech that infants acquire and the most difficult one for adults to modify [5]. Research in this area for monolingual first language acquisition and second language acquisition has unfortunately been rather limited. Studies by Konopczynski [6] and Grabe et al. [2] found that children learning French (traditionally syllable-timed) acquired their speech rhythm earlier than children learning English (traditionally stress-timed). This led Konopczynski [6] to conclude that the acquisition of rhythm is linked to the complexity of rhythmic patterns. Another study [7], which examined the acquisition of rhythm in English, French and German in the productions of four-year-olds led to the same conclusion i.e. that the rhythm (i.e. variability of inter vocalic intervals) of French is acquired earliest followed by German and finally English. Like Konopczynski this study also suggested that there was most likely a correlation between rhythmic complexity and age of acquisition. Thus, children appear to acquire speech rhythm moving from a structurally less variable to an increasingly variable rhythmic pattern. Though all these studies suggest that the syllable timing is probably acquired before stress timing it still remains to be shown when children learning more than one language would exhibit speech rhythms that are characteristic of the language.

The present study is directed towards investigating the question – when do children learning English and Hindi exhibit speech rhythms that are characteristic of the language? To get a better understanding of the rhythmic similarities and dissimilarities of the rhythmic structure of the two languages the children and adults are acquiring, we now provide a short description of the speech rhythms of English and Hindi. The rhythm class hypothesis proposed by Pike [8] and Abercrombie [5] classified spoken languages on the basis of two types of rhythm patterns – (a) stress-timed rhythm and (b) syllable-timed rhythm. According to this hypothesis, speakers of syllable-timed languages exhibit isochronous syllabic intervals wherein syllabic units have equal duration whereas speakers of stress-timed languages exhibit isochronous stress durations wherein the duration from one stressed syllable to the next is believed to be roughly equal. In this context therefore speakers of a syllable-timed language like Hindi would be expected to exhibit a regular pattern of syllabic durations whereas speakers of a stress-timed language, such as English would exhibit a variable pattern of syllable durations. To show this we first obtained spoken language utterances from adult speakers of both English and Hindi. We estimated syllabic durations and variability in syllabic durations for both languages and find that adult speakers of both languages show significantly larger durational variability in English as compared to Hindi. The standard deviation divided by the mean is defined as the durational variability or the co-efficient of variation. To study the relationship between these three parameters we introduce a Variability Correlation Index (VCI). VCI_M is a measure of the correlation between the mean and the variability and VCI_S measures the correlation between the standard deviation and the variability. We find that for both English and Hindi VCI_S shows a strong correlation (> 0.9) as compared to VCI_M (< 0.1) suggesting that the standard deviation plays a major role in the durational variability. We therefore attribute the difference in durational variability between English and Hindi to the difference in the standard deviation of the MSD between English and Hindi.

To investigate when children learning English and Hindi exhibit speech rhythms, which are specific to a language we obtain speech productions in English and Hindi from 70 children. We estimate syllabic durations and calculate durational variability for both languages and find that at 5 years children exhibit similar durational variability for both English and Hindi. However around 7 years of age we find that durational variability for English becomes significantly larger than that of Hindi. We therefore suggest that children learning two languages exhibit syllabic durations that are characteristic of the language at 7 years.

The paper is organized as follows. In Section 2 we describe the participants and database used in the current study and the methods of analysis. In Section 3 we present the results, in Section 4 we discuss conclusions and directions for future research.

2 Participants

Speech samples from 70 children learning both languages in the age group 5–8 years was collected from schools close to New Delhi. The children were classified into four age groups. There were 23 children in the age range from 5.1 to 5.11 (mean age 5.1), 19 children in the age group from 6.1 to 6.11 (mean age 6.1), 16 children who were approximately 7 years (mean age 7.1) and 8 children around 8 years old (mean age 8.1). Eleven adults between the ages of 20–30 also participated in the study. None of the speakers had been diagnosed with a speech, language and auditory problem and all reported normal hearing. They volunteered to participate in the study with parental consent. This study was also approved by the Human Ethics committee of the Institute.

2.1 Recording Setup

A laptop computer along with a simple microphone was used to record the children's speech in Gold wave. Gold Wave is a well featured, screen reader and sound editor. The recordings were done in schools near New Delhi and the adult data was collected from the students and staff of the Institute. We used a close-talking headset microphone to reduce environmental noise, which included traffic (the schools were in an urban area) as well as the school sounds (bells, chairs moving in upstairs classrooms, etc.). Special care was necessary to prevent children from playing with the headset and other cords. The speech was recorded at 22,050 Hz with 16-bit resolution to enable voice source research as well as our recognition studies. The recording sessions with each student were conducted to last less than 15 min so the child would maintain maximum concentration. All participants were administered a picture-naming task which consisted of twenty colored pictures, which included common objects, animals, vehicles. (Examples – pigeon, bus, square, pen, car, apple, ice cream, circle, television, cup, flower, pencil, triangle, plate, scooter, train). For the current analysis only monosyllabic words from English and Hindi were used.

2.2 Data Analysis

The children's productions of monosyllabic words were extracted from the audio recording of the session in consultation with two professional linguists. All recordings where the speech of the volunteer coincided with that of the child were removed. The remaining recordings were phonetically segmented and transcribed using Gold wave. Mean syllabic durations were calculated using code written in Matlab 7.0. The durational variability also called the coefficient of variance (the standard deviation divided by the mean) was the standard measure used to characterize the speech timing patterns. To study the relationship between these three parameters we introduce a Variability Correlation Index (VCI). VCI_M is a measure of the correlation between the mean and the variability and VCI_S measures the correlation between the standard deviation and the variability.

The statistical analyses were carried out using a statistical analysis software, Sigma Stat.

3 Results

3.1 Adults

Syllable durations were calculated from vocal utterances. Figure 1 shows the Mean Syllable Durations (MSD) and Standard Deviation (STDV) in English and Hindi for adults. A plot of the variability in MSD for English and Hindi in Fig. 3.1 shows that the durational variability for English is greater than Hindi. To study the relationship between these three parameters we introduce a Variability Correlation Index (VCI). As seen in Fig. 1 we find that for both English and Hindi VCI_S shows a strong correlation (> 0.9) as compared to VCI_M (< 0.1) suggesting that the standard

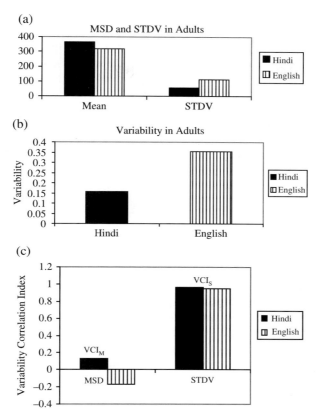

Fig. 1 Mean Syllabic Duration (MSD) and Standard Deviation (STDV) for Hindi and English in adults. (**b**) Variability (standard deviation/mean) of MSD in Hindi and English in adults. (**c**) The Variability Correlation Index (VCI) for MSD and STDV in adults. VCI_M is a measure of the correlation between the mean and the variability. VCI_S measures the correlation between the standard deviation and the variability. For both English and Hindi VCI_S shows a strong correlation as compared to VCI_M, which suggests that the standard deviation plays a major role in the durational variability. In fact for English a negative correlation is seen for VCI_M

deviation plays a major role in the durational variability. We therefore attribute the difference in durational variability between English and Hindi to the difference in the standard deviation of the MSD between English and Hindi.

3.2 Children

To study characteristic speech rhythms in children, we examined variability in syllable duration for English and Hindi as a function of age. As shown in Fig. 2 we see a reduction in variability for Hindi with age whereas in English variability increases with age. For a syllable-timed language (Hindi), the largest values are seen for the

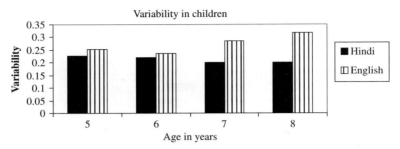

Fig. 2 A comparison of variability (standard deviation/mean) of MSD between English and Hindi for children between 5 and 8 years. For children, rhythmic classification for different languages seems to emerge only around 7 years

youngest children and the smallest value for adults. The reverse trend is seen for English, which is stress-timed.

As seen in Fig. 1, variability in MSD for adults was found to be 0.35 and 0.15 for English and Hindi respectively, indicating smaller variability for Hindi as compared to English. On the other hand for children at ages 5 and 6, there in not much difference in the durational variability for English and Hindi. However by 7 years, we observe a marked increase in the durational variability for English compared to Hindi. A one-way analysis of variance (ANOVA) shows that there is a significant difference in the variability of English as compared to Hindi at 7 years of age ($F(1, 8) = 12.69$, $p < 0.007$). The detail of how this variability emerges is shown in Fig. 3 where we compare the variability in syllabic durations for children from different age groups.

4 Discussion

We examined speech productions in English and Hindi from children between 5 and 8 years to investigate when children learning two languages exhibit characteristic language patterns. Based on the durational variability analysis our results show that at 7 years children exhibit significantly larger variability in syllable duration in English as compared to Hindi. We therefore suggest that children learning two languages exhibit characteristic language rhythms at 7 years. By age 5, as far as durational variability is concerned, the rhythmic patterns produced in English do not differ significantly from those produced in Hindi. Thus as children advance in age they begin to exhibit syllabic patterns that are characteristic of the language. We account for our findings in terms of the greater complexity of rhythmic structure in English as compared to Hindi. As also seen in earlier studies on monolingual children wherein Konopczynski [6] and Grabe et al. [2] found that children learning French (traditionally syllable-timed) acquired their speech rhythm earlier than children learning English (traditionally stress-timed) we find that bilingual children

Speech Rhythms in Children Learning Two Languages 235

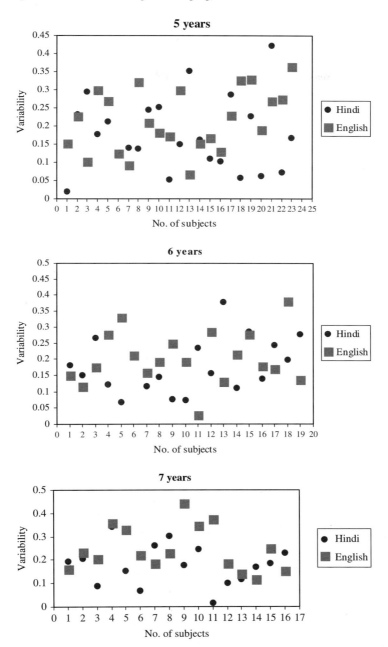

Fig. 3 Figure comparing syllabic variability in English and Hindi for different age groups. As seen for 5-year-old children, there is hardly any difference in the durational variability of English and Hindi. At 6, some differences are seen. However these are not significant. By 7 years, children begin to exhibit increased variability in the syllable durations of English as compared to Hindi and start exhibiting adult-like rhythmic classification for languages

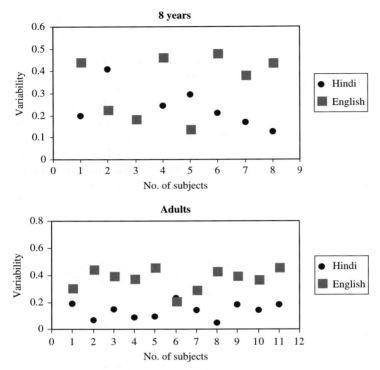

Fig. 3 (continued)

learning a syllable-timed and a stress-timed language acquire syllable-timed patterns earlier than stress-timed patterns.

Our findings are useful for two reasons: this is the first study that shows when children learning two languages belonging to different rhythmic classes exhibit syllabic rhythms specific to the language. It is interesting to note that bilingual children acquire language specific rhythms by 7 years and appear to accomplish this fairly easily. Results from monolingual children learning syllable-timing or stress-timing patterns suggest that children acquire syllable-timed patterns (French) by 4 years while stress-timing is acquired later [2,6]. We were unable to obtain information on when stress-timed rhythm patterns are evident in children. However our data suggests like that of Grabe and Konopczynski that the syllable-timed pattern is acquired earlier than the stress-timed pattern. This shows that we find results, which could have implications in language education at school, wherein the learning of English could be introduced as early as 5 years. Since children have already acquired the syllabic structure of one language and appear capable of learning a second structure which may be more complex without difficulty. Further, this could aid in formulating language syllabi, so that children learning a second language benefit maximum from language training at school. Secondly, development of second language skills is also of clinical interest to speech language pathologists who have been increasingly using psycholinguistic approaches to assess and remedy developmental speech disorders in children. Given the incidence of increase in language disorders among school-going children acquiring a second language [9], it is necessary to obtain data from normally developing children acquiring second language skills to enable clinicians to develop better diagnostic criteria to differentiate between normal and disordered skills. However it is quite clear that more research is required. Our data provide a window into the rhythmic development in two rhythmically diverse languages; complementary data from children younger than 4 are needed, as well as comparable data from other languages. Further, we also observe that children durational variability at 8 years has still not reached adult levels. This could be attributed to a smaller data set at 8 years or it may suggest that children continue to refine speech rhythms beyond 8 years. In future work, we intend to collect such data and carry out a number of studies investigating the development of the rhythm and its acoustic properties at different stages of development. In a country like India with its diversity of languages and cultures this would be interesting and useful.

Acknowledgments This research was funded by the National Brain Research Centre and a research grant from the Department of Information Technology, Government of India. We also wish to thank all the children who participated in the study.

References

1. A. E. Brice, The Hispanic Child: Speech, Language, Culture and Education. Boston, MA: Allyn & Bacon (2002).
2. E. Grabe, U. Gut, B. Post, and I. Watson, The Acquisition of Rhythm in English, French and German. Current Research in Language and Communication: Proceedings of the Child Language Seminar, London: City University (1999).
3. M. Vihman, Phonological Development: The Origins of Language in the Child. Oxford: Blackwell (1996).
4. M. Vihman, R. DePaolis, and B. Davis, Is there a 'trochaic Bias' in Early Word Learning? Evidence from Infant Production in English and French. Child Development, Vol. 69, pp. 935–949 (1998).
5. D. Abercrombie, Elements of General Phonetics. Edinburgh: Edinburgh University Press (1967).
6. G. Konopczynski, A Developmental Model of Acquisition of Rhythmic Patterns: Results from a Cross-Linguistic Study. Proceedings of the International Congress of Phonetic Sciences, Vol. 4, pp. 22–25, Sweden: Stockholm (1995).
7. E. Grabe and E. L. Low, Durational variability in speech and the rhythm class hypothesis. In C. Gussenhoven and N. Warner (Eds.), Papers in Laboratory Phonology, Vol. 7, pp. 377–401, Berlin: Mouton de Gruyter (2002).
8. K. L. Pike, The Intonation of American English. University Press: Michigan (1945).
9. E. Plante and P. M. Beeson, Communication and Communication Disorders: A Clinical Introduction, Boston, MA: Allyn & Bacon (1999).

The Role of Dynamical Instabilities and Fluctuations in Hearing

J. Balakrishnan

Abstract The process of hearing can be understood as one arising through the action of a number of nonlinear elements operating near dynamical instabilities in an environment subject to fluctuations. The sound detector in the inner ear, the mechanoelectrical transducer hair cell can be modelled as a forced Hopf oscillator. When such a system is additionally equipped with a regulatory feedback mechanism which ensures that the system always remains self tuned to operate very close to the bifurcation, then the presence of weak noise can assist in enhancing hugely the amplification of weak stimuli. The fast variable gets phase-locked with the external stimulus for all values of the signal amplitude, showing that the phenomenon is distinct from stochastic resonance. Drawing upon some interesting results obtained for a generic nonlinear system, some speculations can be made in the context of hearing. We suggest a plausible explanation for the hitherto unexplained source of the peaks in the spontaneous otoacoutic emission spectra of various organisms.

Keywords Dynamical instabilities · Fluctuations · Hearing · Nonlinearities · Self-tuning mechanism

1 Introduction

Among the various sensory modalities, hearing constitutes perhaps the most complex one, having evolved over millions of years into an elegant mechanism, integrating several intricately designed subunits, each entrusted with an independent task, and woven beautifully together to work as a single organ in the smallest possible space. Decades of research in hearing have elapsed, yet we have still to fully understand some of the key principles of the micromechanics of hearing.

J. Balakrishnan (✉)
School of Physics, University of Hyderabad, Central University P.O., Gachi Bowli, Hyderabad 500 046, India
e-mail: jbsp@uohyd.ernet.in, janaki05@gmail.com

The sensory cells responsible for detection of sound are the hair cells which lie deep in the inner ear, in the Organ of Corti inside the fluid-filled cochlear duct which runs through the turns of the snail-shaped cochlea. The hair cells are of two kinds – one row of inner hair cells (ihcs) which are the mechanoelectrical transducer sound-detecting cells, and three rows of the outer hair cells (ohcs) which outnumber the ihcs three times over for a reason not known yet. An adult human has about sixteen thousand hair cells in all in a cochlea. The ohcs are piezoelectric in nature and their role in hearing is not completely established since they are present only in mammals. They are believed to play a key role in amplification of the sound detected. Our work is an attempt to understand the generic physical properties underlying the functioning of a single inner hair cell through the study of a general nonlinear dynamical system supporting an oscillatory instability and subject to a fluctuating force. We regard the ihc as a nonlinear element operating close to a dynamical instability.

2 The Mechanotransducer Hair Cell and Nonlinearities

The ear is a remarkably sensitive sensor of audio signal. A perfect signal detector, the ear operates over a wide range of frequencies and intensities, and is endowed with the ability to selectively respond to frequencies, to detect and amplify very weak stimuli, and to adapt to sustained stimuli. A normal ear also is intensively active, emitting spontaneously sounds of its own, a phenomenon known as spontaneous otoacoustic emission. To understand the dynamical features exhibited by the ear, it has been modelled since long as a nonlinear element (see refs. [1–10] for example, for some few early papers).

Recent experiments have revealed deeper insights into the process of sound detection at the cellular scale. These have led to some theoretical advances in hearing research [2–22]. The inner hair cell consists of a cell body, on top of which is pivoted a bundle of some 50–200 ciliated structures (microvilli), arranged in a particular order of ascending height. Each ciliated structure is called a stereocilium and their ensemble crowning every hair cell is called the hair bundle. The top of each stereocilium is linked to the side of the next adjacent higher stereocilium by means of a thin filamentous structure called the tip-link (Fig. 1). Mechanically gated ion channels are located at these attachment points on the sides of the stereocilia. Each stereocilium comprises of several actin filaments encased by a plasma membrane. The opening and closing of the ion channels is accomplished through the binding and unbinding of proteins at the terminal ends of the tip links with a group of channel motor proteins (myosin) which move up and down the actin filaments in the stereocilia. This happens at a very rapid rate, at about a thousand times a second [23, 24].

The hair bundles are immersed in the endolymph fluid of the cochlear duct since they project out of the top surface of the Organ of Corti which rests upon the basilar membrane. At the bottom basolateral surface the hair cell makes synapses with axons which join with the auditory nerve.

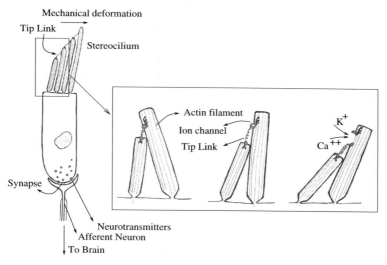

Fig. 1 Mechanoelectrical transduction by an inner hair cell. Inset shows the traditional gating spring model: deflection of the hair bundle causes the stereocilia to bend and produces a shear between each of them. This produces a tightening of the tip links between them; the channels open in response to the tension in the tip links enabling influx of ions into the hair cell through the stereocilia

Sound incident upon the ear is collected by the pinna and transmitted via the middle ear where it is detected. The vibrations of the eardrum are transferred through the three smallest bones of the body, the ossicles, which drum on the oval window of the cochlea. The resulting pressure gradient in the endolymph fluid and the up and down motion of the basilar membrane produce a deflection of the hair bundle which moves together as a unit because the stereocilia are linked together. This produces a shearing motion between the stereocilia which stretches the tip-links producing a tension in them. This in turn pulls open the gates of the ion channels, letting in an influx of ions such as Ca^{2+}, K^+, Mg^{2+}, etc, into the hair cell through the stereocilia (Fig. 1). The influx causes a change in the cell membrane potential which triggers the release of neurotransmitters through the synapses into the afferent nerves, carrying the signal through the auditory nerve to the brain. Thus sound is detected.

The hair cells are arranged all along the basilar membrane in a specific tonotopic arrangement. The ones present closer, nearer the basal end of the cochlea have shorter and stiffer hair bundles and they detect high frequency sounds while those at the apical end detect low frequency sounds and their hair bundles are longer and floppier. The mechanical properties of the hair bundle are responsible to a large measure for its characteristic frequency.

Compressive nonlinearity in the response of the ear results in the amplification of low input sounds while leaving high input sounds relatively unamplified. In fact, the ear responds to vibrations of atomic dimensions emanating from extremely faint signals [25]. To explain the amplification of such low intensity stimuli, it is believed that the ears contain active force generating elements. Such active elements, if present, would also account for the phenomenon of spontaneous otoacoustic emissions from the ears of all organisms.

The system of the single hair cell too presents all the features of a nonlinear system operating close to an instability – a highly nonlinear response of the hair bundle deflection, spontaneous deflections of the hair bundle in the absence of any stimulus, sharp sensitivity and ability to amplify weak inputs, and selective response to a particular characteristic frequency.

In recent years therefore there has emerged a stream of literature [26–29] modelling the hair cell with the hair bundle deflections as a nonlinear oscillator capable of producing self-sustained oscillations – a Hopf oscillator. Spontaneous hair bundle oscillations could then be ascribed to self-sustained oscillatory behaviour of the Hopf oscillator at the bifurcation point. Such an oscillator operating at the edge of the instability would also show a large gain for a weak input [26–29].

Based on these ideas, Magnasco and coworkers [26–28] did a careful detailed analysis of a biophysical model and showed that their hair bundle model permitted a Hopf instability, and suggested that such a model would require a regulatory mechanism to ensure that the operating point of the system was always near the bifurcation point. Careful, controlled experiments by Hudspeth and coworkers [30] demonstrated that the hair bundle's spontaneous oscillations were indeed those of a Hopf oscillator.

Following the work of [26–28], Camalet et al. [29] discussed the same problem for hair bundle deflections in more generic terms, emphasizing the need to admit a regulatory self-tuning mechanism to keep the hair cell poised all the time close to the Hopf bifurcation, but the mechanism of self-tuning was not discussed by them.

As the hair bundle is immersed in the endolymph fluid, its motion is subject to drag forces due to the fluid viscosity; its environment is also noisy. It is therefore surprising how the hair bundle achieves its end of being able to sense and amplify very faint signals. Jaramillo & Wiesenfeld suggested [31] that brownian motion of the hair bundle amplifies limitingly low amplitude stimuli through the phenomenon of stochastic resonance. They based their explanation on the gating spring model [23, 24] for mechanoelectrical transduction in which the opening of a mechanoreceptive ion channel shortens the associated gating spring, lowering its tension and also the energy of the open state in comparison to the closed state. Their model however does not explain the spontaneous oscillations of the hair bundles, nor does it explain the high frequency selectivity of each hair cell. Therefore it has been of interest to understand the effect of fluctuations on a hair cell modelled as a Hopf oscillator, and to study how a self-tuning regulatory mechanism can work in the presence of noise. A mathematical study of feedback tuning for a hair cell in the absence of noise was discussed in [32, 33] and in [34], two adaptation mechanisms were used to demonstrate computationally that the hair cell can produce "self-sustained critical oscillations".

In [35] we studied how a regulatory feedback mechanism for "self tuning" a generic nonlinear system to always operate close to the Hopf instability, would work in the presence of both external periodic forcing as well as additive noise. We calculated the distance of the operating point from the bifurcation for the generic system as a measure of the tuning which could possibly be achieved. The details of that calculation are presented in [35] and will not be discussed here. We interpreted these results in the biological context of the hair cell system.

3 Generic Hopf Oscillator with Feedback and the Hair Cell

Active hair bundle motion has been modelled in recent literature [29] in the following manner. Let $x(t)$ denote the instantaneous hair bundle deflection in response to a stimulus F_{ext} and $y(t)$ denote the active force generated by the active component of the hair bundle, arising from the motion of the channel motor protein (myosin) complex along an actin filament in the stereocilium. The linear behaviour of the hair bundle displacement is then described by the coupled equations:

$$\lambda \frac{dx}{dt} = -kx + y + F_{\text{ext}}(t)$$
$$\beta \frac{dy}{dt} = -y - \bar{k}x \qquad (1)$$

λ, k and β stand respectively for the drag coefficient, stiffness of the hair bundle and relaxation time of the active process and $\bar{k}$ has dimensions of a spring constant. The nonlinear terms (not shown in (1)) arise from the gating spring model for the mechanoreceptive ion channels and depend upon the channel opening probability P_{open} which is a sigmoidal function of x. In the absence of F_{ext}, linear stability analysis shows that the system admits a Hopf bifurcation in a certain operating range [26–29]. We have studied such a set of equations admitting a Hopf bifurcation, in the most generic form, the normal form, since that gives us greater insight into the physical processes at work and also it gives us a certain amount of predictive power. The aim is to capture the most salient features of the hair bundle system by a study of such a normal form, so as to attempt to understand thereby why nature's design of the sound sensor is the best possible one. For simplicity we assume that the externally applied stimulus is sinusoidal in nature, of a single frequency Ω. The generic normal form equation for the system (1) in the presence of an external periodic forcing $F_0 e^{i\Omega t}$ is [36]:

$$\frac{dz}{dt} = A(\omega, C)z - B(\omega, C)|z|^2 z + O(|z|^4 z) + F_0 e^{i\alpha_0} e^{i\Omega t} \bar{z}^{s-1} \qquad (2)$$

where A and B are complex coefficients which depend upon a control parameter C, and ω denotes the characteristic frequency of the unforced Hopf oscillator. For a hair bundle, α_0 depends upon its mechanical properties. Equation (2) describes a system which is in $s:m$ resonance (s, m being mutually coprime): $\Omega = \omega \left(\frac{s}{m} + \gamma\right)$ where γ denotes the detuning parameter.

We have studied the simplest situation of 1:1 resonance, with a vanishing detuning parameter. We take the control parameter C to be the concentration of the Ca^{2+} ions entering the stereocilia. This is in accordance with the experimental finding of Lumpkin and Hudspeth [37, 38] that Ca^{2+} plays a regulatory role in the mechanotransduction process. Since the hair bundle deflection controls the opening and closing of the gates of the ion channels, which in turn dictates the amount of calcium entering the cell, we have a situation in which the control parameter C depends upon one of the system variables $x(t)$. C regulates the operating point of the hair cell system, always bringing it close to the bifurcation point.

Therefore we consider the most general regulatory feedback equation where the control parameter for the generic system (2) is dependent on one of the system variables

$$x(t): \frac{dC}{dt} = \Gamma\big(C, x(t)\big) \qquad (3)$$

$\Gamma(C, x(t))$ is, in general, a nonlinear function of C.

A bifurcation involves a sudden qualitative change in the dynamical behaviour of the system as a parameter changes and at this point, the nature of the equilibrium changes [30]. On one side of the critical value of the control parameter the system is quiescent and upon crossing the critical value, it becomes oscillatory.

For a periodic force $F_{ext}(t)$, $x(t)$ can be expanded in terms of its normal modes through a Fourier expansion. Since C depends upon x, it can be expanded in a Fourier series also, and it is possible to find a solution of (2) in terms of its normal modes [29, 35]. Assuming that the first mode is the dominant one, we find that in the close proximity of the bifurcation, and for a vanishingly small force,

$$f_1 \approx \chi_1(C_0)x_1 - \chi_2(C_0)|x_1|^2 x_1 \qquad (4)$$

where $\chi_1(C_0)$ and $\chi_2(C_0)$ are some functions of C_0 and f_1 is the first term in the expansion of F_{ext}. Very close to the critical point, we see that the first term goes to zero, the term nonlinear in x takes over, and the response varies as one third power of the input. The implication of this result is that small inputs are amplified hugely, the gain of the system being given by

$$\text{Gain} = \frac{\text{Output}}{\text{Input}} = f_1^{-2/3} \qquad (5)$$

Thus a nonlinear device can in principle be used as a good signal detector if its operating point is tuned always to be as close as possible to a Hopf bifurcation. The self-sustained oscillatory behaviour associated with the birth of the limit cycle could then model the spontaneous oscillations of the hair bundle – setting the force to zero in (4), this can be expressed as [29, 35]:

$$|x_1| \approx \Delta \left(\frac{C_c - C}{C_c}\right)^{1/2}, \qquad (6)$$

where the ratio $\frac{\chi_1(C_0)}{\chi_2(C_0)}$ has been expressed in a form incorporating the known scaling behaviour [30] of the amplitude of the limit cycle with the distance $(C_c - C)$ from the bifurcation point C_c. Here Δ denotes a characteristic saturating value for the x variable.

The compressive nonlinearity of the hair bundle response seen in the power law in (5) is indeed observed experimentally (see [39]) indicating that the Hopf oscillator model for the hair bundle is a good one.

4 A Generic Hopf Oscillator in the Presence of Noise

To understand the key features of hair bundle behaviour in its physical noisy environment, we considered a generic Hopf oscillator subject to additive noise. Introducing a complex Gaussian white noise $\xi(t) = \xi_1(t) + i\xi_2(t)$, with noise correlations defined by:

$$\begin{aligned}
<\xi_1(t)\xi_1(t')> &= Q_a \delta(t-t') \\
<\xi_2(t)\xi_2(t')> &= Q_b \delta(t-t') \\
<\xi_1(t)\xi_2(t')> &= Q_{ab} \delta(t-t'),
\end{aligned} \qquad (7)$$

we obtain the Langevin equations

$$\frac{dz}{dt} = f(z, \bar{z}, t) + \varepsilon^{1/2} \xi(t) \qquad (8)$$

where $f(z, \bar{z}, t)$ denotes the terms on the right hand side of (2).

In the presence of noise, the bifurcation point is smeared out. Our aim was to understand how well a physical Hopf oscillator in noisy surroundings would act as detector of weak signals: how close to the deterministic bifurcation point can we actually bring the system, using an in-built self-tuning feedback mechanism?

We address this question by first rewriting (8) in a form more suitable for analysis. We Taylor expand A and B around C_c and combine (6) with the Fourier expansion of x, keeping only the dominant mode x_1. Then using the polar coordinate representation of $z = r(t)e^{i\phi t}$, (8) can be rewritten as

$$\dot{r} = \beta r - \left(l - \beta'\frac{C_c}{\Delta^2}\frac{\cos^2\phi(t)}{4\cos^2(\Omega t + \alpha)}\right)r^3$$
$$+ F_0 \cos(\Omega t + \alpha_0 - \phi(t)) + O(r^5) + \alpha_1\xi_r$$
$$\dot{\phi} = -\omega(C_c) - \left(-d(C_c) + \omega'(C_c)\frac{C_c}{\Delta^2}\frac{\cos^2\phi(t)}{4\cos^2(\Omega t + \alpha)}\right)r^2$$
$$+ F_0 \frac{\sin(\Omega t + \alpha_0 - \phi(t))}{r} + O(r^4) + \frac{\alpha_2\xi_\theta}{r} \qquad (9)$$

where

$$\alpha_1\xi_r = (\cos\phi\,\xi_1 + \sin\phi\,\xi_2)\varepsilon^{1/2}$$
$$\alpha_2\xi_\theta = (\cos\phi\,\xi_2 - \sin\phi\,\xi_1)\varepsilon^{1/2} \qquad (10)$$

and

$$\beta = Re(A(C_c)); \quad l = Re(B(C_c)); \quad \beta'(C_c) = Re(A'(C_c))$$
$$\omega = -Im(A(C_c)); \quad d = -Im(B(C_c)); \quad \omega'(C_c) = -Im(A'(C_c)) \qquad (11)$$

We have introduced a smallness parameter ε (the inverse of the extensivity parameter V which denotes the system size) for the noise. Using Stratonovich calculus, we find that the equivalent Fokker-Planck equation equivalent to (17) is time dependent:

$$\frac{\partial P(r,\phi,t)}{\partial t} = -\frac{\partial}{\partial r}\left[\beta r - lr^3 + F_0\cos(\Omega t + \alpha_0 - \phi)\right.$$
$$\left. + \beta'\frac{C_c}{\Delta^2}\frac{\cos^2\phi(t)}{4\cos^2(\Omega t + \alpha)}r^3 + \frac{\varepsilon}{2r}Q_{\phi\phi}\right]P(r,\phi,t)$$
$$-\frac{\partial}{\partial \phi}\left[-\omega + dr^2 + \frac{F_0}{r}\sin(\Omega t + \alpha_0 - \phi)\right.$$
$$\left. -\omega'\frac{C_c}{\Delta^2}\frac{\cos^2\phi(t)}{4\cos^2(\Omega t + \alpha)}r^2 - \frac{\varepsilon}{r^2}Q_{r\phi}\right]P(r,\phi,t)$$
$$+\frac{\varepsilon}{2}\left[\frac{\partial^2}{\partial r^2}Q_{rr} + \frac{\partial^2}{\partial \phi^2}\frac{Q_{\phi\phi}}{r^2} + 2\frac{\partial^2}{\partial r\partial \phi}\frac{Q_{r\phi}}{r}\right]P(r,\phi,t) \qquad (12)$$

where

$$\frac{1}{2}Q_{rr} = Q_a\cos^2\phi + 2Q_{ab}\sin\phi\cos\phi + Q_b\sin^2\phi$$
$$\frac{1}{2}Q_{r\phi} = -Q_a\cos\phi\sin\phi + Q_{ab}(\cos^2\phi - \sin^2\phi) + Q_b\sin\phi\cos\phi$$
$$\frac{1}{2}Q_{\phi\phi} = Q_a\sin^2\phi - 2Q_{ab}\sin\phi\cos\phi + Q_b\cos^2\phi \qquad (13)$$

Extending the phase space through the transformation: $\lambda = \Omega t + \alpha$ enables us to rewrite (12) in an autonomous form.

The physical quantities of interest such as the response of the system to a stimulus, and the distance from the bifurcation which gives the extent to which feedback tuning can be effected, are obtained after performing noise averages, and to calculate these, we need to first determine the form of the probability distribution near the bifurcation.

In order to study the effect of the fluctuations at the onset of the limit cycle, we use the singular perturbation technique following the elegant work of [40, 41]. Details of this may be found in [35]. We find that the radial variable exhibits amplified non-Gaussian fluctuations on a slow time scale. This makes it possible to factor the probability for the full system as the product of a time dependent part for the radial distribution, and a time independent part for the fast variable ϕ:

$$P(r, \phi, \lambda, t) = P(\phi | r, \lambda) P(r, \lambda, t) \tag{14}$$

To extract the dynamics of the physically interesting center radial modes, we perform an averaging over the fast variable ϕ. We find that the term containing the external forcing vanishes unless $\phi = \lambda \pm 2n\pi$. Thus there occurs necessarily a phase locking of the fast variable with the external forcing frequency.

In the context of the hair bundle, this result implies that the motion of the active component (the myosin-5 motor protein) in the stereocilia gets phase-locked with the external stimulus frequency.

Details of the analysis and the calculations can be found in reference [35], here we reproduce only the results. We obtain the following analytical solution of the Fokker-Planck equation (12) in the adiabatic limit, for small frequencies:

$$\lim_{t \to \infty} P_{ad}(r, \lambda, t) = N_e r \exp\left\{ -\frac{2}{Q\varepsilon} \left[\left(1 - \frac{\mu \beta' C_c}{8\Delta^2 \cos^2 \lambda}\right) \frac{r^4}{4} - \beta \frac{r^2}{2} - \nu F_0 r \right] \right\} \tag{15}$$

where the normalization constant N_e is given by

$$N_e = N_0 \left\{ \sum_{n=0}^{\infty} \frac{\Gamma\left(\frac{n}{2} + \frac{3}{2}\right)}{n!} \left(\frac{2\nu F_0}{(Q\varepsilon)^{\frac{3}{4}} \left(1 - \frac{\mu \beta' C_c}{8\Delta^2 \cos^2 \lambda}\right)^{\frac{1}{4}}} \right)^n \right.$$

$$\left. D_{-\frac{n}{2} - \frac{3}{2}} \left(\frac{-\beta}{\left[Q\varepsilon \left(1 - \frac{\mu \beta' C_c}{8\Delta^2 \cos^2 \lambda}\right)\right]^{\frac{1}{2}}} \right) \right\}^{-1}, \tag{16}$$

and the constants ν and μ take the values

$$\nu = \begin{cases} 1 & \text{for} \quad \phi(t) = \Omega t + \alpha \pm 2n\pi \\ 0 & \text{otherwise} \end{cases}$$

$$\mu = \begin{cases} 2\cos^2\lambda & \text{when} \quad \nu = 1 \\ 1 & \text{otherwise.} \end{cases}$$

In (15), (16),

$$Q = Q_a + Q_b. \tag{17}$$

$$N_0 = 4\pi \left(\frac{\left(1 - \frac{\mu\beta'C_c}{8\Delta^2\cos^2\lambda}\right)}{Q\varepsilon} \right)^{\frac{3}{4}} \exp\left[-\frac{\beta^2}{4Q\varepsilon\left(1 - \frac{\mu\beta'C_c}{8\Delta^2\cos^2\lambda}\right)} \right] \tag{18}$$

and D_{-n} are parabolic cylinder functions.

For the forced system the long time limit of the probability distribution is stationary because the fast variable over which averaging is performed is phase-locked with the external driving frequency and rotates with it. On the other hand in the case of the unforced system we find that the probability distribution as $t \to \infty$ is time dependent – this comes about because of the presence of a feedback of the form in (3) which makes the system explicitly non-autonomous.

Quantities of practical interest such as the response of the system can now be calculated. In the limit of weak noise, the response is found following [42–45] by expanding $< x(t, \alpha) >_{ad}$ in a Fourier series:

$$< x(t, \alpha) >_{ad} = \sum_{-\infty}^{\infty} M_n e^{in(\Omega t + \alpha)} \approx 2|M_1|\cos(\Omega t + \alpha) \tag{19}$$

Figure 2A and B plot the response as a function of the noise strength. At certain very small values of the noise, we see that the system amplifies its response enormously. This is also seen in Fig. 3 which plots the spectral power amplification η_{as} as a function of noise for different signal strengths. We find that a noisy Hopf oscillator with feedback control amplifies weaker signals better. This behaviour is similar to the observed response of the ear which responds more to weaker signals. Indeed, it is known that cochlear amplification is most pronounced at the auditory threshold and falls steeply as the intensity of the stimulus increases [22, 26–28]. Therefore noise assists the system in amplifying enormously, weak signals, but the phenomenon appears to be different from stochastic resonance because in our case there is phase-locked behaviour of the fast variables with the external frequency for all values of the signal amplitude.

Our results agree qualitatively with known experimental facts about the micromechanics of the ears of various organisms. Figure 4 shows the experimental findings of Ruggero & coworkers [46] demonstrating the larger response of the basilar

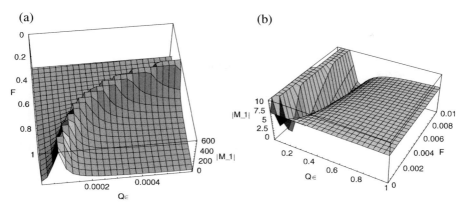

Fig. 2 (a) 3D plots of the response $|M_1|$ vs Q_ε, (b) F_0 of resonantly forced self-tuned Hopf-oscillator in the presence of additive noise Q_ε (for $\beta = 0.16$, $l - \frac{\beta' C_c}{4\Delta^2} = 0.76$)

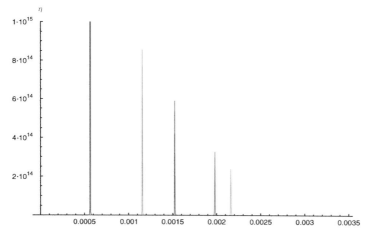

Fig. 3 Spectral power amplification η of a resonantly forced self-tuned Hopf oscillator as a function of the noise strength Q_ε (for $\beta = 0.16$, $l - \frac{\beta' C_c}{4\Delta^2} = 30.02$; from left to right, $F_0 = 0.0025$, 0.005, 0.0065, 0.0083, 0.009.)

membrane for weaker stimuli, at any given frequency. The constructive role of noise in the auditory system for amplifying weak signals was elucidated by Jaramillo & Wiesenfeld in [31], wherein they suggested that the hair cell employed the Brownian motion of its hair bundle to amplify very weak stimuli through stochastic resonance. Their work demonstrated conclusively the noise dependence of mechanoelectrical transduction. Figure 5 reproduced from [31] plots the signal to noise ratios (SNRs) as a function of hair bundle noise in mechanoelectrical transduction by the hair cells of the frog sacculus stimulated at 500 Hz. The SNR shows enhanced values at lower noise strengths.

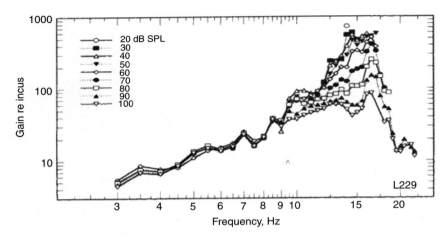

Fig. 4 Response of the basilar membrane in the hook region, 1.5 mm from the end, of the chinchilla cochlea (as reported by: [46])

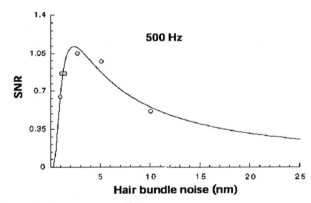

Fig. 5 Signal to noise ratio of mechanoelectrical transduction as a function of noise added to the stimulus probe for a hair cell stimulated at 500 Hz [31]

The experimentally observed features in Figs. 4 and 5 are captured in the behavior of a resonantly forced noisy Hopf oscillator plotted in Fig. 3 not only are weaker signals amplified more, but also for a given signal strength, amplification and response are better at lower noise strengths. While the model in [31] is unable to explain spontaneous motion of the hair bundle, our work in [35] gives the extent to which the feedback mechanism of the system tunes the hair bundle to operate in the proximity of the region of instability making possible spontaneous self-sustained oscillations, and exhibiting noise-assisted amplification of weak signals through a mechanism akin to, but different from stochastic resonance.

The nature of the feedback (3) causes the self tuned unforced system ($\nu = 0$, $\mu = 1$ in (15,16)) to be time-periodic and hence one expects it to exhibit the characteristic precursors of the Hopf instability [47, 48]. With this in view, we calculate the asymptotic spectral density $S_{as}(\psi)$ of the unforced system from the phase-averaged asymptotic correlation function $\overline{K}_{as}(t,t') = <<x(t)x(t')>>_\alpha$.

We find [35] that to lowest order approximation,

$$\overline{S}_{as}(\psi) = \int_{-\infty}^{\infty} d\tau \overline{K}_{as}(\tau) e^{-i\psi\tau}$$
$$\approx \int_{-\infty}^{\infty} d\tau \frac{1}{8\pi} \left(\frac{Q\varepsilon}{2\beta} + h(\tau) \right) e^{-i(\psi-n\omega)\tau}. \quad (20)$$

Here, $\tau = t - t'$ and the time periodicity of $\overline{K}_{as}(\tau)$ arising from the feedback (the self-tuning terms) is used to expand it in Fourier series. The time-dependent periodic part has been represented by $h(t) = h(t+T)$. We see that the spectrum consists of delta peaks at frequencies $(\psi - n\omega)$, $n = 0, 1, 2, \ldots$ superimposed on the spectrum of a bounded time-periodic function. The periodicity arises because of the oscillatory nature of the feedback. Figure 6 plots $\overline{S}_{as}(\psi)$ for the unforced system for a given noise strength [49]. The spectrum we obtain in Fig. 6 is similar to that of spontaneous otoacoustic emissions recorded from the ears of all organisms (see [50–52]). For comparison, in Fig. 7 we reproduce from the work of Stuart & Hudspeth [53] and Göpfert et al. [54], the SOAE arising from the hearing organs of a lizard and a fruit fly respectively. As in Fig. 6, the experimental plots generally have

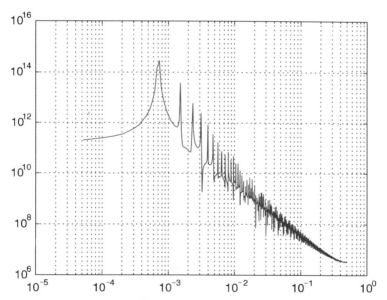

Fig. 6 Asymptotic spectral density $\overline{S}_{as}(\psi)$ of an unforced noisy Hopf oscillator with feedback tuning (20) as a function of frequency ψ (for $\beta = 0.01$, $C_c/\Delta^2 = 110.1$, $\Omega = \omega = 0.5$, $\alpha = \alpha_0 = 0.2$)

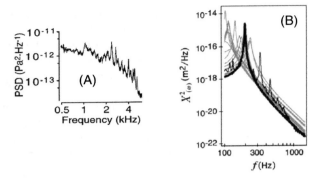

Fig. 7 Otoacoustic emission spectra: (**a**) Power spectral density of spontaneous otoacoustic emissions from a ear with total emission power of $700\,\mu Pa^2$, of the Tokay gecko, as a function of frequency [53]. (**b**) Power spectrum of the sound receiver in the antennal hearing organ of a live wild type *Drosophila melanogaster* fly. The self-sustained receiver oscillations give an estimate of the energy contributed by the motile mechanosensory neurons to which they are connected [54]

an enhanced maximum about a particular frequency and exhibit peaks throughout at various frequencies. These observations led us in [35] to provide a possible explanation for the occurrence of peaks in the SOAE spectra. We suggested that they originate because of the presence of a feedback mechanism of the kind in (3) which manifests via a noisy precursor of the Hopf instability. In other words, there is a manifestation of the phenomenon of coherence resonance [47, 48] in the unforced noisy Hopf oscillator.

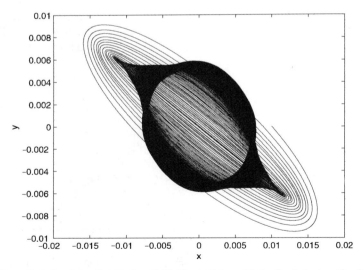

Fig. 8 Phase plot of a deterministic forced Hopf oscillator with feedback tuning (for $\beta = 0.01$, $C_c/\Delta^2 = 30.1$, $\Omega = \omega = 0.5$, $\alpha = \alpha_0 = 0.02$)

The actual biological details of the particular auditory system under study (such as the mechanical properties of the hair bundle, the length of the basilar membrane, the influence of various attachments of the hair bundle and the effect of coupling of a hair cell Hopf oscillator with the others via the basilar membrane, etc.) would give a more accurate and realistic determination of the spectral density and determine the slope of the SOAE plot.

The phase plot of a forced Hopf oscillator with feedback control shows interesting structure for certain values of the parameters [55]. An example of such a situation is shown in Fig. 8. In the context of hearing, one could interpret the phase plot to reflect the dynamics driving the external stimulus to pull in the hair bundle system into a phase-locked situation corresponding to the true limit cycle attractor after a certain initial transient time interval. A detailed analysis and possible explanation is presented in [55].

5 Conclusion

Modelling a single mechanoelectrical transducer hair cell of the ear as a system close to a Hopf instability can explain much of its observed nonlinear behaviour. We studied a generic nonlinear system and found that fluctuations play a major role in hugely amplifying weak stimuli, weaker signals being amplified better – a behaviour similar to that shown by the ear. The noise-assisted amplification mechanism seems to be a phenomenon different from stochastic resonance as the system's fast variable is entrained with the applied frequency for all values of the signal amplitude.

Modelling physiological systems using generic nonlinear models can be helpful in getting a deeper insight into the physical mechanisms at work. The bounds on the amount of tuning which can be achieved with a general feedback system of the form in (3) [55] provide a useful measure for building sensitive detectors, and in the audio range, for emulating the behaviour of the ear – the best audio sensor provided by Nature.

References

1. T. Gold, *Proc. R. Soc. B*, **135**, 492 (1948).
2. J. Zwislocki, *J. Acoust. Soc. Am.*, **22**, 778 (1950).
3. O. F. Ranke, *J. Acoust. Soc. Am.*, **22**, 772 (1950).
4. T. J. Goblick, Jr., & R. R. Pfeiffer, *J. Acoust. Soc. Am.*, **46**, 924 (1969).
5. H. Duifhuis, *J. Acoust. Soc. Am.*, **59**, 408 (1976).
6. E. Zwicker, *Biol. Cybern.*, **35**, 243 (1979).
7. C. R. Steele & L. A. Taber, *J. Acoust. Soc. Am.*, **65**, 1001 (1979).
8. D. O. Kim, C. E. Molnar & J. W. Mathews, *J. Acoust. Soc. Am.*, **67**, 1704 (1980).
9. S. T. Neely, *J. Acoust. Soc. Am.*, **78**, 345 (1985).
10. E. Zwicker, *J. Acoust. Soc. Am.*, **80**, 146 (1986).
11. R. A. Eatock, D. P. Corey & A. J. Hudspeth, *J. Neurosci.*, **7**, 2821 (1987).
12. A. J. Hudspeth, *Nature*, **341**, 397 (1989).
13. J. A. Assad, N. Hacohen & D. P. Corey, *PNAS*, **86**, 2918 (1989).

14. P. Dallos, *J. Neurosci.*, **12**, 4575 (1992)
15. J. A. Assad & D. P. Corey, *J. Neurosci.*, **12**, 3291 (1992)
16. A. J. Hudspeth & P. G. Gillespie, *Neuron*, **12**, 1 (1994)
17. A. C. Crawford & R. Fettiplace, *J. Physiol.*, **312**, 377 (1981)
18. G. A. Manley, *J. Neurophysiol.*, **86**, 541 (2001)
19. A. J. Hudspeth, Y. Choe, A. D. Mehta & P. Martin, *PNAS*, **97**, 11765 (2000)
20. F. Jaramillo, V. S. Markin & A. J. Hudspeth, *Nature*, **364**, 527 (1993)
21. D. P. Corey & A. J. Hudspeth, *J. Neurosci.*, **3**, 962 (1983)
22. M. A. Ruggero, *Curr. Opin. Neurobiol.*, **2**, 449 (1992)
23. J. Howard & A. J. Hudspeth, *PNAS*, **84**, 3064 (1987)
24. J. Howard & A. J. Hudspeth, *Neuron*, **1**, 189 (1988)
25. W. Bialek, *Ann. Rev. Biophys. Biophys. Chem.*, **16**, 455 (1987)
26. Y. Choe, M. O. Magnasco & A. J. Hudspeth, *Proc. Natl. Acad. Sci. USA*, **95**, 15321 (1998)
27. V.M. Eguiluz, M. Ospeck, Y. Choe, A.J. Hudspeth & M.O. Magnasco, *Phys. Rev. Lett.*, **84**, 5232 (2000)
28. M. Ospeck, V. M. Eguiluz & M. O. Magnasco, *Biophys. J.*, **80**, 2597 (2001)
29. S. Camalet, T. Duke, F. Jülicher & J. Prost, *PNAS*, **97**, 3183 (2000)
30. J. Guckenheimer & P. Holmes, *Nonlinear Oscillations, Dynamical Systems & Bifurcations of Vector Fields*, Springer-Verlag, (1983)
31. F. Jaramillo & K. Wiesenfeld, *Nature Neurosci.*, **1**, 384 (1998)
32. L. Moreau & E. Sontag, *Phys. Rev.*, **E 68**, 020901(R) (2003)
33. L. Moreau, E. Sontag & M. Arcak, *Syst. Control Lett.*, **50**, 229 (2003)
34. A. Vilfan & T. Duke, *Biophys. J.*, **85**, 191 (2003)
35. J. Balakrishnan, *J. Phys. A: Math. Gen.*, **38**, 1627 (2005)
36. C. Hemming & R. Kapral, *Faraday Discuss.*, **120**, 371 (2001)
37. E. A. Lumpkin & A. J. Hudspeth, *PNAS*, **92**, 10297 (1995)
38. E. A. Lumpkin & A. J. Hudspeth, *J. Neurosci.*, **18**, 6300 (1998)
39. P. Martin & J. Hudspeth, *PNAS*, **96**, 14306 (1999)
40. C. Van den Broeck, M. Malek Mansour & F. Baras, *J. Stat. Phys.*, **28**, 557 (1982)
41. F. Baras, M. Malek Mansour & C. Van den Broeck, *J. Stat. Phys.*, **28**, 577 (1982)
42. P. Jung & P. Hänggi, *Europhys. Lett.* **8**, 505 (1989)
43. P. Jung & P. Hänggi, *Phys. Rev.* **A 41**, 2977 (1990)
44. L. Gammaitoni, P. Hänggi, P. Jung & F. Marchesoni, *Rev. Mod. Phys.*, **70**, 223 (1998)
45. P. Jung & P. Hänggi, *Phys. Rev.* **A 44**, 8032 (1991)
46. S. S. Narayan & M. A. Ruggero, in *Proceedings of the Symposium on Recent Development in Auditory Mechanics*, eds. H. Wada, T. Takasaka, K. Ikeda, K. Ohyama, & T. Koike (World Scientific Publishing, US, UK, Singapore, 2000)
47. K. Wiesenfeld, *J. Stat. Phys.*, **38**, 1071 (1985)
48. A. Neiman, P. I. Saparin & L. Stone, *Phys. Rev.* **E 56**, 270 (1997)
49. B. Ashok & J. Balakrishnan, (submitted) (2008)
50. P. M. Zurek, *J. Acoust. Soc. Am.*, **69**, 514 (1981)
51. W. Denk & W. W. Webb, *Hear. Res.*, **60**, 89 (1992)
52. C. Köppl, in *Advances in Hearing Research*, ed. G.A. Manley, C. Köppl, H. Fastl & H. Oeckinghaus, pp. 200–209 (World Scientific, Singapore, 1995
53. C. E. Stuart & A. J. Hudspeth, *PNAS*, **97**, 454 (2000)
54. M. C. Göpfert, A. D. L. Humphris, J. T. Albert, D. Robert & O. Hendrich, *PNAS*, **102**, 325 (2005).
55. B. Ashok, J. Balakrishnan & G. Ananthakrishna (in preparation) (2008)

Electrical Noise in Cells, Membranes and Neurons

Subhendu Ghosh, Anindita Bhattacharjee, Jyotirmoy Banerjee, Smarajit Manna, Naveen K. Bhatraju, Mahendra. K. Verma and Mrinal K. Das

Abstract Fluctuation analysis has been an important aspect of research in various disciplines e.g. physics, chemistry, geology, environmental sciences, biology and others. The fluctuation in a system is either the Noise created by external forces or due to the internal dynamics. Interestingly, many of these systems exhibit $1/f$ power spectrum despite large differences in their basic properties. In physiology, a good number of reports related to noise have come up during recent years. Advanced electrophysiological experiments, as reported earlier, enable us to record currents through a single or a group of channels on a cell or a lipid bilayer membrane. We discuss the time series behaviour of the channel currents and the associated noise. The role of noise in a living system is not very well understood. Nevertheless, the quantification of noise at the ion channel level has thrown light on the phenomenon of transport of ions and metabolites across cell membrane and its mechanisms. Power Spectrum analysis of current indicates powerlaw noise of $1/f$ nature. We discuss the origin of $1/f$ noise in open ion channels. The process is recognized as a phenomenon of self-organized-criticality (SOC) like sand pile avalanche and other physical systems. For multi-channels, Power Spectral Density, we found, is a good parameter to probe collective behavior of ion channels. Neuronal communications and transfer of action potential in neurons through synapses has been recognized to be the key parameter for functioning of the brain. We demonstrate through computational approach that synaptic noise, a kind of channel noise, enables weak input signals in the neurons to evoke electrical spikes, thus plays an important role in the process of brain functioning.

Keywords Ion channel · Voltage dependent anion channel · Gaussian noise · Single neuron · Limit cycle · Spike train · Synchronization

---

S. Ghosh (✉)
Department of Biophysics, University of Delhi South Campus, Benito Juarez Road, New Delhi 110021, India
e-mail: profsubhendu@gmail.com

1 Introduction

Noise analysis has been an important tool in understanding the dynamics of systems belonging to a wide range of disciplines, e.g. physics, chemistry, geology, environmental sciences, biology and others. In physiology, a good number of reports related to noise have come up during recent years [1]. Advanced electrophysiological techniques, enable us to record current noise, both stationary and non-stationary, through a single or a group of ion channels on a cell or a lipid bilayer membrane [2, 3]. Development on quantification of noise at the ion channel level has lead to newer understanding of the mechanism of transport of ions and metabolites across cell membrane [4, 5]. In brain the communications among the neurons are extremely imortant and this is known to take place via transfer of action potentials or electric pulses through synapse, a neuron-to-neuron junctional device comprising of ion channels [6]. Synaptic noise, a kind of channel noise, plays an important role in this process [1, 6, 7]. Here we present an overview of the role of noise in ion channels in cell or lipid bilayer membranes and neurons.

In the first part of this paper we discuss the properties of current noise in ion channels, in specific, Voltage Dependent Anion Channel (VDAC), and a theoretical model to explain the noise pattern. This is done with the following background. Rostovtseva and Bezrukov have studied the noise in VDAC during ATP transport [4] and in OmpF during PEG transport [5], and reported white noise. Bezrukov and Winterhalter [8] observed $1/f$ noise for similar porin channels. There are reports on experimental evidence and analysis of open channel noise (stationary) by Mak and Webb in alamethicin [9], by Hainsworth et al. in K^+ channels from sarcoplasmic reticulum [10], and by Zhou et al. in CFTR channel pore [11]. In addition, there are reports on non-stationary noise [12].

In the second part we explore the role of noise in the electrical activity of a neuron, e.g. spiking. Action Potentials or Electrical Spikes are known to be the language of neuronal communication [13]. These action potentials are the result of inward and outward currents that pass through the ion channels, namely Na^+ and K^+ channels respectively, coupled with the capacitance in the cell membrane, over a time scale. When a stimulus is input in a neuron it accumulates the electrical charge till it crosses the threshold value [13–15] beyond which an output signal or Action Potential or spike is generated and transmitted to the adjacent neuron. The entire process takes place in a collective manner in a large number of neurons. As expected strong inputs are capable of evoking spikes by overcoming the threshold, whereas weak signals are not [16, 17]. Interestingly, weak signals can also evoke spikes when coupled with noise [16, 18–21]. Here we discuss some results on the aforesaid role of noise in a single neuron receiving weak input stimulus. This is based on the spatio-temporal equation (nonlinear) of the action potential, the Hodgkin Huxley's equation. The investigations are fully computational.

2 Noise in VDAC

VDAC is an abundant protein in the outer mitochondrial membrane, which forms large voltage gated pore in planar lipid bilayers, and act as the pathway for the movement of substances in and out of the mitochondria by passive diffusion [22,23]. VDAC essentially plays an important role in the transport of ATP, ions, and other metabolites between the mitochondrion and the cytoplasm. Importantly, VDAC has been reported to be responsible for cytochrome C mediated programmed cell death or apoptosis [24–27]. All VDACs form channel with roughly similar single channel conductance (4.1 nS to 4.5 nS in 1 M KCl). This voltage gated pore has an effective diameter 2.5 nm to 3 nm and develops cation preference in closed or lower conductance states [28–32]. In our previous papers the gating behaviour of VDAC has been discussed along with the existence of its capacitance [33, 34]. VDAC has been reported to exist in the outermembrane of brain cells [35] and its specific role in cell surface is under exploration. Here, we look into the nature of current noise (stationary) through VDAC as observed in the experiments.

2.1 The Experiment

VDAC was purified from rat brain mitochondria using the method of De Pinto et al. [36], and reconstituted into the planar lipid bilayers according to the method of Roos et al. [37]. Aqueous compartments on both sides of the bilayer membrane are connected to an integrating patch amplifier Axopatch 200 A (Axon Instruments, USA) through a matched pair of Ag/AgCl electrodes. Axopatch 200A was connected to an IBM computer through an interface Digidata 1322A (Axon Instruments, USA). Channel current due to transport of K^+ and Cl^- was recorded using the data acquisition software Clampex (pClamp 9.0, Axon Instruments, USA). The experiment was performed on anti-vibration table (TMC, USA) to avoid any vibrational noise. Single channel recording of VDAC was performed in a symmetric bath solution.

We have recorded ion current arising from single-channel in presence of externally applied potential across the membrane. Data were filtered at 200 Hz using 8-pole Bessel Filter and sampled at a frequency (1 kHz) greater than the corner frequency with an ITC-16 interface. A time series consists of open and closed states. Here, we focus only on open states, whose typical time trace for +25 mV is shown in Fig. 1.

2.2 Data Analysis and Results

The following noise characteristics have been studied: (i) probability distribution of large current amplitudes ($P(I)$ vs. I), (ii) probability distribution function of inter

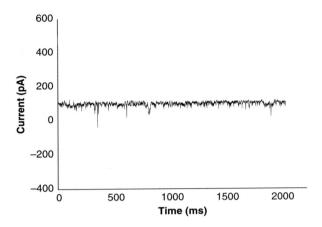

Fig. 1 Continuous current trace (open state) of rat brain VDAC at +25 mV. Membrane bathing solution consisted of 500 mM KCl, 10 mM Hepes, and 5 mM MgCl$_2$. Its pH value was 7.4. The experiment was done at room temperature (23–25°C)

burst intervals (between two consecutive current bursts), and (iii) power spectral density $S(f)$.

(i) To compute the probability distribution of large current amplitudes $P(I)$, we have taken time traces for open states with voltage of -25 mV, and obtained dataset of 241764 points. Now we plot a histogram of $P(I)$ vs. I in log-log scale, which is shown in Fig. 2. We find that $P(I) \propto I^{-a}$ with $a = 5.0 \pm 0.28$. Hence, the amplitudes of large fluctuations obey a power law. This result is similar to the avalanche size or earthquake size distribution, and it clearly indicates that the transport in the ion-channel is a non-equilibrium process. It may be noted that the equilibrium processes have typically Gaussian probability distribution.

(ii) Now we compute the statistics of time-intervals between two successive large current fluctuations in current traces at a constant volatge (+25 mV and +20 mV). We take single time trace and compute the standard deviation of the fluctuations. The large current fluctuations are marked by taking noise signals which are three times the standard-deviation. After the large current fluctuations have been identified, we measure the time-interval τ between two consecutive large current fluctuations. We find that $P(\tau) \simeq \tau^{-\beta}$ with $\beta = 1.9 \pm 0.1$ at +25 mV and 1.8 ± 0.2 at +20 mV. Similar analysis for different current cutoffs are performed. Results show that the level of $P(\tau)$ changes with cutoff, but the slopes are approximately the same (within error bar). We also notice that the single channel traces with low channel conductance and closed states do not follow any powerlaw.

(iii) Computation of power spectral density $S(f)$ was done using the software Clampfit. We used 2064-point vectors. The plot of spectral density $S(f)$ versus frequency f for a current time trace of open VDAC channel at +25 mV is shown in Fig. 3. It is evident from the figure that powerlaw $f^{-\alpha}$ fits quite well to $S(f)$ vs. f plot for frequency range of more than two decades. For the experimental voltages (± 30 mV) the slope ranges from 0.72 to 1.05. These results indicate that the noise exhibits $1/f$ spectrum which is consistent with

Fig. 2 Log-log plot of probability distribution of magnitude of current fluctuation $P(I)$ vs. I of VDAC at -25 mV. The data size is 24176 points

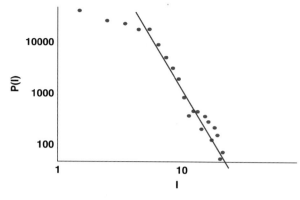

Fig. 3 The plot of power spectral density $S(f)$ vs. frequency f for a full open state of VDAC at $+25$ mV. A power law $1/f^\alpha$, $\alpha = -0.92$, fits the data quite well

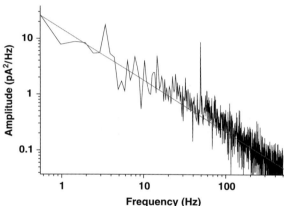

that of Bezrukov and Winterhalter [8]. The positivity of spectral index ($\alpha > 0$) implies that noise is correlated at large-scales, and correlation lengths are large. The divergence of correlation length is referred to as critical behaviour for the system. Interestingly, ion channels exhibit powerlaw behaviour for the noise under steady state without any fine tuning.

2.3 A Model for 1/f Noise in Ion Channel

Theoretical understanding of powerlaw noise in ion-channels is in its infancy. Earlier, Bezrukov and Winterhalter [8] argued that $1/f$ noise in ion-channel current is not a fundamental property of non-equilibrium transport phenomena, rather it reflects the complex hierarchy of equilibrium protein dynamics. Siwy et al. have discussed the aforesaid problem in biological and synthetic channels on lipid bilayer membrane and claimed that the powerlaw noise (non-stationary) originates from the channel's opening and closing [12]. They analyzed the channel gating in view of

Markov process and tested through Smoluchowski-Chapman-Kolmogorov (SCK) equation. In maltoporin channel the cause of $1/f$ noise in current has been thought to be due to collisional friction experienced by maltodextrin sugar molecules passing through the channel [38].

In general, systems under equilibrium have thermal noise, which are typically uncorrelated or white. Hence noise in ion-channels (mostly $1/f$) is non-thermal, and therefore, correlated. Also, the transport process through ion-channel is non-equilibrium or driven because the current is directed due to external potential. Keeping these in view, we provide a theoretical model for ion-channel noise. Earlier, scientists have proposed various schemes to explain $1/f$ noise, e.g. activated random processes, diffusion [39], self-organized criticality (SOC), etc. The non-equilibrium nature of ion-channel transport rules out diffusion. The transport in ion-channel is a steady-state process, and it cannot be modeled satisfactorily by relaxation process. Hence, activated-random process appears to be ruled out as well. Considering that the transport in ion-channel is a non-equilibrium steady-state flow, self-organized criticality (SOC) appears to be the most promising candidate to explain $1/f$ noise in ion-channel [40]. That is the reason why we have focused on SOC in this work.

Bak et al. proposed self-organised criticality as the source of $1/f$ noise in sand-pile avalanches [41, 42]. They illustrated their theory using random avalanches in sand-pile as events. Sand is poured slowly at random position of the sand-pile. At critical conditions, avalanches of various sizes and duration occur in the system. Here we draw an analogy between the sand pile avalanche and ion transport through channels. We claim that the random obstruction of ions during the passage through an ion channel follows SOC dynamics.

In the following, we propose a novel mechanism to explain $1/f$ noise in ion channel. Our derivation is inspired by a model for intermittency route to chaos [43]. Recently some interesting studies have been reported which show that the waiting-time or the first-time return probability between earthquakes of magnitude M or greater has a powerlaw distribution, i.e., $P(\tau) \simeq \tau^{-\beta}$, where τ is the time gap between two earthquakes of magnitude M or greater and β is the exponent [44, 45]. The correlation is built up by the accumulation of stresses over time.

To argue for the self-organized-criticality of the VDAC channel dynamics it was shown that the first-time return probability of large fluctuation of current in the time series exhibit powerlaw distribution, i.e., the probability of occurence of large (three times the standard deviation or greater) magnitude of current flutuations obeys a powerlaw, i.e.,

$$P(\tau) \simeq \tau^{-\beta} \qquad (1)$$

It may be noted that the shape of ion channel is highly dynamic; the irregular inner surface of the channel act as barriers. It is reasonable to expect that dynamical variation of pore-width can induce correlations because ions pile up near the barrier. The large fluctuations arise because the ions pass through the pore intermittently. So depending on the size of the ion pile, large current fluctuations occur at a very short interval of time as well as long interval. The above-mentioned powerlaw distribution for the first-time return may be due to variable width.

In conclusion, we propose that the noisy patterns in the current fluctuations in open VDAC channel is due to random but correlated obstruction of ions during the passage through the channel. Powerlaw distribution indicates the phenomenon of self-organized criticality. This self-organized criticality is the cause of $1/f$ noise in open state current through VDAC. In an earlier publication we have reported that the aforesaid $1/f$ noise of VDAC gets converted to white noise while interacting with various ligands, e.g. Bax, plasminogen etc. [46]. This is attributed to the decrease in dynamics and temporal correlation due to ligand binding. Multi-channel noises are yet to be studied in detail. We found multi-channel stationary noise can be an important parameter to probe channel-channel cooperativity.

2.4 Universal Scaling Law on the Basis of Noise Analysis

As mentioned above there is a phenomenological commonality between ion transport through ion channels and the sand-pile avalanche, solar flares, recurrence of earthquake [47]. All these processes are thought to be following SOC. In sand-pile avalanche, earthquake there exist scaling laws. We found that similar scaling law holds for ion channels, e.g. VDAC. This motivates us to propose a Universal Scaling Law [48].

$$P(I_c, \tau) = I^{-\alpha} \tau^{-\beta} g(\tau/\tau_c) \qquad (2)$$

where, g is a universal function and c refers to criticality.

3 Noise in a Single Neuron

Hodgkin and Huxley (H-H) proposed a model for the propagation of Action Potential based on the results of their experiment on giant squid nerve axon [49]. Following Hodgkin and Huxley's (HH) equation a number of models describing Action Potential or Neuronal Spike have been proposed [13, 17].

3.1 The Hodgkin-Huxley Model

We consider the four-dimensional Hodgkin-Huxley model of action potential [49] as follows:

$$C\frac{dV}{dt} = I - \overline{g_{Na}}m^3 h(V - V_{Na}) - \overline{g_K}n^4(V - V_K) - \overline{g_L}(V - V_L) \qquad (3a)$$

$$\frac{dn}{dt} = \alpha_n(V)(1-n) - \beta_n(V)n \qquad (3b)$$

$$\frac{dm}{dt} = \alpha_m(V)(1-m) - \beta_m(V)m \qquad (3c)$$

$$\frac{dh}{dt} = \alpha_h(V)(1-h) - \beta_h(V)h \qquad (3d)$$

where

$$\alpha_n(V) = \frac{0.01(V+55)}{1 - \exp[-(V+55)/10]},$$

$$\beta_n(V) = 0.125 \exp[-(V+65)/80]$$

$$\alpha_m(V) = \frac{0.1(V+40)}{1 - \exp[-(V+40)/10]},$$

$$\beta_m(V) = 4 \exp[-(V+65)/18]$$

$$\alpha_h(V) = 0.07 \exp[-(V+65)/20]$$

$$\beta_h(V) = \frac{1}{1 + \exp[-(V+35)/10]}$$

where V is the membrane potential and I is the externally applied current in the neuron by some electrode, m, n, h are the dimensionless gating variables whose values lie in the interval [0, 1]. C is the membrane capacitance. $\alpha_i(V)$ and $\beta_i(V)$ are the rate constants of ion transport from inside to outside and from outside to inside respectively.

Initial conditions required to solve this system of equations may be obtained by assuming that the membrane potential is at the same value for a long time. In this case n, m and h assume their steady state values [49] given by

$$\frac{\alpha_n}{\alpha_n + \beta_n}, \; m_\infty = \frac{\alpha_m}{\alpha_m + \beta_m} \; \text{and} \; h_\infty = \frac{\alpha_h}{\alpha_h + \beta_h} \qquad (4)$$

The constants of H-H equation are as follows:-
C = $1 \mu F/cm^2$, V_{Na} = 50 mV, V_K = −77 mV, V_L = −54.4 mV, $\bar{g}_{Na}$ = 120 mmho/cm², $\bar{g}_K$ = 36 mmho/cm², $\bar{g}_L$ = 0.3 mmho/cm².

We reduce the four dimensional H-H equation to 2D by considering $m = m_\infty$ as m evolves faster than n and h. It is observed that the periodic action potential corresponds to the functional relationship between sodium and potassium activation i.e. $h(t) + n(t) \approx 0.8$. Therefore the reduced 2D H-H equations are:

$$\frac{dV}{dt} = [I - \bar{g}_{Na}\{m_\infty(V)\}^3(0.8-n)(V-V_{Na}) - \bar{g}_K n^4(V-V_K) - \bar{g}_L(V-V_L)]/C \qquad (5a)$$

$$\frac{dn}{dt} = \alpha_n(V)(1-n) - \beta_n(V)n \qquad (5b)$$

For a noisy input a noise term is added to the right hand side of equation (5a)

$$\frac{dV}{dt} = [I - \overline{g_{Na}}\{m_\infty(V)\}^3(0.8-n)(V-V_{Na}) - \overline{g_K}n^4(V-V_K)$$
$$- \overline{g_L}(V-V_L)]/C + D.\xi(t) \qquad (6)$$

3.2 Computational Method

For numerical integration we have used MATLAB 6.1 and obtained the phase portrait, time series, power spectrum using the values of the parameters mentioned in the previous section. We have explored numerically the dynamics of Hodgkin-Huxley model using the externally applied input current. In order to solve non-linear equations Runge-Kutta method of fourth order has been applied. Firstly, the original model has been simulated. This has been followed by the simulation of a model which is subjected to noisy inputs. The time series thus generated as results are used for drawing phase portrait, power spectrum etc. In the noise induced system the neuron receives a noisy input defined by Gaussian function $\xi(t)$ such that

$$<\xi(t)>= 0 \text{ and } <\xi(t)\xi(t')>= D^2\delta(t-t') \qquad (7)$$

where D is the strength of the noise. MATLAB function RANDN is called for generating the random noise in case of noisy input. We have also used chi-square test to analyze the Inter Spike Interval distribution.

3.3 Numerical Results

In this work our aim is to show the role of Gaussian noise in the continuous generation of action potential or spike train. Here, we have dealt with HH model of action potential and the Gaussian noise. So far, we have discussed the cases of (i) constant (DC), (ii) periodic and (iii) impulse signal as the applied input current. For a noise free system with cases (i) and (iii) we find the spike solution followed by a damped oscillation when the input is below the critical level. However in case (ii), the oscillatory output is recurring but of very low amplitude, given the experimental parameters. The recurring action potentials are generated after the input crosses a critical value. This is an expected result from HH and other models. In addition, our computational results suggest that instead of increasing the external input if a noisy system receives low input and the noise strength is adequate then it exhibits continuous oscillatory output, which is not possible in a noise free system (Fig. 4). In other words, the noise intensity after crossing the critical value compensates for

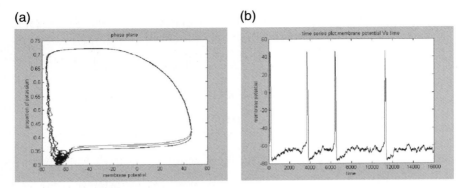

Fig. 4 (a) Phase plane for Spike and Noisy input given by $I = A\delta(t - n\tau)$ where A is the Amplitude, t is time, τ is the period of occurrence of impulse and $n = 1, 2, 3, 4, 5, \ldots$, $A = 10$ mAmp, $\tau = 0.1$ msec, Gaussian Noise Strength $D = 20$. (b) Time Series corresponding to (a)

a weak signal in order to evoke infinite oscillations or spikes. We conclude that the low input current drives the system to periodic oscillation provided the input current is noisy.

During the entire computation we have looked into the quantitative relation between input current amplitude A and the threshold strength of noise (D_{TH}) required for evoking spike train. Spike and Noisy input is given by $I = A\delta(t - n\tau)$, t is time, τ is the period of occurrence of impulse (= 0.1 msec) and $n = 1, 2, 3, 4, 5, \ldots$. We find in general D_{TH} as the parabolic function of the current amplitude A as shown by the following relations:

$$D_{TH} = 20 - 0.2514A - 0.0057A^2 \qquad (8)$$

In a living state neurons receive stimuli externally as well as from other neurons, and these could be both strong and weak. While strong inputs are capable of evoking action potentials or spike trains in the neurons, weak signals are required to be deciphered by the neurons in order to retrieve more information. We believe the latter could be achieved by coupling with noise or fluctuation, which exists in the neuronal system.

4 Noise in Ensemble of Neurons and Brain Functioning

The case of single neuron although reveals lot of interesting features related to brain dynamics, it is an isolated and rather artificial situation. In fact, in the functioning of the brain the most important factor is the interactions and communications within an ensemble of neurons. It is being argued that during a number of physiological processes the spiking trains of neurons get synchronized [6, 50]. Zhou and Kurths have demonstrated (numerically) that Gaussian noise enhances synchronization of weakly coupled neurons and coherence of the spike trains [51]. The effects of noise on the synchronization in a model system of neuronal assembly have been investigated by McMillen and Kopell [52], Fernandez et al. [53] and Wang et al. [54]. However, the experimental support of the above-mentioned effect of noise is awaited. If established this will throw a new light on the functioning of the brain.

References

1. J. A. White, J. T. Rubenstein, and A. R. Kay, *Channel noise in neurons*, Trends. Neurosci. **23** 131–137 (2000)
2. K. Benndorf, *In: Single-channel recording*, eds. B. Sakmann, and E. Neher, Plenum Publishing, New York, USA 129–145 (1995)
3. R. K. Adair, *Noise and stochastic resonance in voltage-gated ion channels*, Proc. Natl. Acad. Sci. **100** (21) 12099–12104 (2003)
4. T. K. Rostovtseva, and S. M. Bezrukov, *ATP transport through a single mitochondrial channel*, Biophys. J. **74** (5) 1865–1873 (1998)
5. T. K. Rostovtseva, E. M. Nestorovich, and S. M. Bezrukov, *Partitioning of differently sized Poly(ethyleneglycol)s into OmpF porin* Biophys. J. **82** 110–115 (2002)
6. H. Haken, *Brain dynamics*, Springer **13** 54–73 (2002)
7. L. M. Ward, *Dynamical cognitive neuroscience*, MIT Press, Cambridge, Massachusetts, London, England 135–153 (2002)
8. S. M. Bezrukov, and M. Winterhalter, *Examining noise level at the single-molecule level: 1/f noise of an open maltoporin channel*, Phys. Rev. Lett. **85** 152–155 (2000)
9. D. O. Mak, and W. W. Webb, *Molecular dynamics of alamethicin transmembrane channels from open- channel current noise analysis*, Biophys. J. **69** 2337–2349 (1995)
10. A. H. Hainsworth, R. A. Levis, and R. S. Eisenberg, *Origins of open-channel noise in the large potassium channel of sarcoplasmic reticulum*, J. Gen. Physiol. **104** 857–883 (1994)
11. Z. Zhou, S. Hu, and T. C. Hwang, *Voltage-dependent flickery block of an open cystic fibrosis transmembrane conductance regulator (CFTR) channel pore*, J. Physiol. **532** (2) 435–448 (2001)
12. Z. Siwy, and A. Fulinsky, *Origin of $1/f^\alpha$ noise in membrane channel currents*, Phys. Rev. Lett. **89** 108101 (2002)
13. J. G. Dumas and A. Rondepierre, *Modelling the Electrical activity of a neuron by a continuous and pecewise affine hybrid system: Computation and control*, Springer, New York (2003)
14. H. C. Silvia, C. M. Luciana and M. E. S. Reneto, *Brain and mind fundamentals, how brain cells work*, State University of Campinas, Brazil (2000)
15. P. Nelson, *Biological physics*, W. H. Freeman & Co., New York 508 (2004)
16. Y. Yu, W. Wang, J. Wang, and F. Liu, *Resonance enhanced signal detection and transduction in the Hodgkin-Huxley neuronal systems*, Phys. Rev. E. **63** 021907 (2001)
17. A. Scott, *Neuroscience: A mathematical primer*, Springer, New York (2002)
18. C. Heneghan, C. C. Chow, J. J. Collins, T. T. Imhoff, S. B. Lowen, and M. C. Teich, *Information measures quantifying aperiodic stochastic resonance*, Phys. Rev. E. **54** 2228–2231 (1996)
19. F. Moss, and K. Wiesenfeld, *Benefits of background noise*, Scientific American, 66–69 (1995)
20. J. K. Douglass, L. Wilkens, E. Pantazelou, and F. Moss, *Noise enhancement of information transfer in crayfish mechanoreceptors by stochastic resonance*, Nature, **365** 337–340 (1993)
21. S. Wang, F. Liu, and W. Wang, *Impact of spatially correlated noise on neuronal firing*, Phys. Rev. E. **69** 011909 (2004)
22. M. Colombini, *Anion channels in the mitochondrial outer membrane*, Curr. Top. Membr. **42** 73–101 (1994)
23. V. De Pinto, A. Messina, R. Accardi, R. Aiello, F. Guarino, M. F. Tomasello, M. Tommasino, G. Tasco, R. Casadio, R. Benz, F. De Giorgi, F. Ichas, M. Baker, and A. Lawen, *New functions of an old protein: the eukaryotic porin or voltage dependent anion selective channel (VDAC)*, Ital. J. Biochem. **52** (1) 17–24 (2003)

24. J. Banerjee, and S. Ghosh, *Interaction of mitochondrial voltage-dependent anion channel from rat brain with plasminogen protein leads to partial closure of the channel*, Biochim. Biophys. Acta **1663** 6–8 (2004)
25. J. Banerjee, and S. Ghosh, *Phosphorylation of rat brain mitochondrial voltage-dependent anion channel as a potential tool to control leakage of cytochrome c*, J. Neurochem. **98** 670–676 (2006)
26. S. Shimizu, Y. Matsuoka, Y. Shinohara, Y. Yoneda, and Y. Tsujimoto, *Essential role of voltage- dependent anion channel in various forms of apoptosis in mammalian cells*, J. Cell. Biol. **152** (2) 237–250 (2001)
27. T. K. Rostovtseva, W. Tan, and M. Colombini, *On the role of VDAC in apoptosis: Fact and fiction*, J. Bioenerg. Biomembr. **37** 129–142 (2005)
28. R. Benz, *Structure and function of porins from gram-negative bacteria*, Annu. Rev. Microbial. **42** 359–393 (1988)
29. O. Ludwig, J. Krause, R. Hay and R. Benz, *Purification and characterization of the pore forming protein of yeast mitochondrial outer membrane*, Eur. Biophys. J. **15** 269–276 (1988)
30. R. Benz, M. Kottke, and D. Brdiczka, *The cationically selective state of the mitochondrial outer membrane pore: a study with intact mitochondria and reconstituted mitochondrial porin*, Biochim. Biophys. Acta. **1022** 311 (1990)
31. O. S. Smart, J. Breed, G. R. Smith, and M. S. Sansom, *A novel method for structure-based prediction of ion channel conductance properties*, Biophys. J. **72** 1109–1126 (1997)
32. M. Colombini, E. Blachy-Dyson and M. Forte, in *Ion channels*, Volume 4, edited by T. Narahashi (Plenum Publishing Corp., New York), 169–202 (1996)
33. S. Ghosh, A. K. Bera, and S. Das, *Evidence for nonlinear capacitance in biomembrane channel system*, J. Theoretic. Biol. **200** (3) 299–302 (1999)
34. S. Ghosh, and A. K. Bera, *Role of H+ concentration on the capacitance of a membrane channel*, J. Theoretic. Biol. **208** 383–384 (2001)
35. E. A. Jonas, J. Buchanan, and L. K. Kackzmarec, *Prolonged actvation of mitochondrial conductances during synaptic transmission*, Science **286** (5443), 1347–1350 (1999)
36. V. De Pinto, G. Prezioso, and F. Palmieri, *A simple and rapid method for purification of the mitochondrial porin from mammalian tissues*, Biochim. Biophys. Acta. **905** 499–502 (1987)
37. N. Roos, R. Benz, and D. Brdiczka, *Identification and characterization of pore-forming protein in the outer membrane of rat liver mitochondria*, Biochim. Biophys. Acta **686** 204–214 (1982)
38. R. Dutzler, T. Schirmer, M. Karplus, and S. Fischer, *Translocation mechanism of long sugar chains across the maltoporin membrane channel*, Structure **10** 1273 (2002)
39. P. Dutta, and P. M. Horn, *Low-frequency fluctuations in solids: 1/f noise*, Rev. Mod. Phy. **53**, 497–511 (1981)
40. E. Milotti, *1/f noise: a pedagogical review*, arXiv/Physics/0204033 (2002)
41. P. Bak, C. Tang, and K. Wiesenfeld, *Self-organized criticality: An explanation of the 1/f noise*, Phys. Rev. Lett. **59** 381–384 (1987)
42. P. Bak, C. Tang, and K. Wiesenfeld, *Self-organized criticality*, Phys. Rev. A. **38** 314–374 (1988)
43. H. G. Schuster, and W. Just, *Deterministic chaos: an introduction*, John Wiley and Sons, New York, USA (2005)
44. P. Bak, K. Christensen, L. Danon, and T. Scalon, *Unified scaling law for earthquakes*, Phys. Rev. Lett. **88** 123001–123004 (2002)
45. X. Yang, S. Du, and J. Ma, *Do earthquakes exhibit self-organized criticality?* Phys. Rev. Lett. **92** 178501–178504 (2004)
46. J. Banerjee, and S. Ghosh, *Investigating interaction of ligands with voltage dependent anion channel through noise analyses*. Arch. Biochem. Biophys. **435** 369–371 (2005)
47. J. Banerjee, M. K. Verma, S. Manna, and S. Ghosh, *Self organized criticality and 1/f noise in single-channel current of voltage dependent anion channel*, Europhys. Lett. **73** 1–7 (2006)

48. M. Verma, S. Manna, J. Banerjee, and S. Ghosh, *Universal scaling laws for large events in driven non-equilibrium systems*, Europhys. Lett. **76** 1050–1056 (2006)
49. J. Moehlis, *Canards for a reduction of the Hodgkin-Huxley equations*, J. Math. Biol. **52** 141–153 (2005)
50. L. Glass, *Synchronization and rhythmic processes in physiology*, Nature **410** 277–284 (2001)
51. C. Zhou, and J. Kurths, *Noise induced synchronization and coherence resonance of a Hodgkin–Huxley model of thermally sensitive neurons*, Chaos **13** 401–409 (2003)
52. D. McMillen, and N. Kopell, *Noise stabilized long-distance synchronization in population of model neurons*, J. Comput. Neurosci. **15** 143–157 (2003)
53. L. F. L. Fernandez, F. J. Corbacho, and R. Huerta, *Connection topology dependence of synchronization of neural assemblies on Class 1 and 2 excitability*, Neural Networks **14** 637–696 (2001)
54. Y. Wang, D. T. W. Chik, and Z. D.Wang, *Coherence resonance and noise induced syncronization in globally coupled Hodgkin-Huxley neurons*, Phys. Rev. E. **61**, 740–746 (2000)

Index

A

Attractor, 3–17, 23–27, 186, 194, 253

B

Baroreflex sensitivity (BRS), 34, 155, 157, 161, 163
Behavior, 8, 22, 23, 27, 70, 85, 102, 104, 107, 108, 112, 118, 124, 125, 131
Bilingual, 229, 230, 234, 236
Blood pressure variability, 33–35, 159
Breathing, 33–39, 41–45, 156, 167, 171

C

Cardiac arrhythmias, 35, 51–54, 63, 69, 90
Cardiac fibrillation, 69, 90
Cardiac surgery, 155, 163, 165
Cardiac tissue, 51, 53, 55, 63, 69, 70, 77, 79, 86, 89–91, 102, 104, 107, 108, 125, 128
Cardiorespiratory synchrogram (CRS), 167–169, 172, 173, 176–179
Causality, 33, 34
Chaos, 28, 31, 81, 86, 104, 107, 108, 125, 185–188, 193, 195, 260
  See also Spatiotemporal chaos
Chaos control, 69, 70, 73, 78, 80
Chaos suppression, 94, 95, 102, 104, 107
Children, 171, 229–237
Concentric waves, 89, 107
Congestive heart failure (CHF), 139, 146–148, 150, 152
Correlation, 10, 16, 21–24, 29, 33, 35, 37–45, 91, 140, 141, 143, 145, 146, 151, 169, 175, 177, 179, 186, 194, 206, 207, 209, 230–233, 245, 251, 259–261
Courtship language, 215
Criticality, 139, 140, 146, 147, 150, 255, 260, 261

D

Delay coordinates, 3
Dynamical instabilities, 239–253

E

ECG, 21, 22, 27, 30, 31, 36, 52, 135, 148, 149, 156, 171, 173
EEG, 21, 27–31, 185, 195, 201, 207, 210
Embedding, 3–7, 11, 13–15, 17, 23, 25, 193, 194
Empirical mode decomposition, 167–180
English, 229–236
Epilepsy, 22, 185–210
Excitable media, 51, 53, 69–77, 80, 91, 92, 95, 104, 108, 118

F

Feedback decoupling, 185, 204, 208, 209
Fluctuations, 27, 29, 45, 57, 130, 139, 141–143, 145–147, 150–152, 157, 158, 239–253, 255, 258–261, 264

G

Gaussian noise, 75, 140, 142, 143, 145, 205, 255, 263, 264
Generalized dimension, 21, 22, 24

H

Hearing, 231, 239–253
Heart rate variability (HRV), 33–35, 127, 134, 139–152, 155, 157–159, 161, 163
Heart rate variability models, 127
Hindi, 229–236

I

Ictogenesis, 185, 204, 210
Intermittency, 135, 139, 140, 147, 150, 152, 171, 179, 260
Intrinsic mode functions (IMFs), 167–172, 179

269

I

Ion channel, 51, 52, 55, 64, 71, 72, 86, 127, 128, 134, 135, 240–244, 255, 256, 258–261

L

Language rhythm, 229, 230, 234
Limit cycle, 110, 131, 133, 244, 245, 247, 253, 255
Low-energy defibrillation, 76, 89, 90, 104

M

Markov model, 215, 219, 223
Multifractal spectrum, 21

N

Non-Gaussianity, 139, 140, 150, 152
Nonlinear oscillations, 29, 127–135, 242
Nonlinearities, 22, 27, 31, 71, 99, 130, 204, 239–241, 245

P

Postoperative autonomic regulation, 155

S

Self-tuning mechanism, 239, 242
Single neuron, 255, 256, 261, 264
Spatial synchronization of chaos, 185

Spatiotemporal chaos, 51, 53, 54, 56, 69, 70, 72–80, 85, 90–94, 96, 104
Spike train, 255, 263, 264
Spiral turbulence (ST), 51–53, 55–60, 62, 63, 69–86
Spiral waves, 51–63, 69, 72, 73, 75, 76, 81, 89, 90, 92–95, 97, 100–102, 104, 107, 108, 118, 124, 125
Syllable duration, 229, 230, 232–235
Symbolic dynamics, 22, 159, 215
Synchronization, 107–125, 129, 167–179, 185, 187–192, 195–203, 206, 207, 209, 210, 255, 264, 265

T

Time series, 3–5, 7–10, 12–14, 16, 17, 21–27, 29, 31, 36, 56–61, 140, 142, 150–152, 157, 159, 168–170, 172, 175–177, 179, 193, 215, 217, 255, 257, 260, 263, 264
Time series analysis, 21, 45

V

Voltage dependent anion channel (VDAC), 255–261

W

Wave propagation, 53, 89